Oklahoma LANDSCAPES
A Century of Change

by
T. H. Milby and Forrest L. Johnson

Edited by Bob Burke

Oklahoma Horizons Series

SERIES EDITOR: GINI MOORE CAMPBELL
ASSOCIATE EDITOR: ERIC DABNEY

OKLAHOMA LANDSCAPES

Printed in the United States of America by
Baker Group, LLC & Jostens, Inc. - 405.503.3207

ISBN 978-1-885596-66-6

Library of Congress
Catalog Number 2008921979

Designed by Kris Vculek

OKLAHOMA HERITAGE ASSOCIATION

OKLAHOMA CITY, OKLAHOMA
888.501.2059
www.oklahomaheritage.com

CONTENTS

Foreword

Paul G. Risser

Former Oklahoma Chancellor of Higher Education

This book offers a revealing view of our state that allows us to integrate space and time, people and places, motivations, and consequences. Photographs taken at the end of the 1800s and early 1900s depict a state that is both the same and different from today. Sorting out the sameness, as well as the differences, requires introspection at various levels of intensity and rewards us with both comfort and discomfort.

At the most general level, we are struck with the marvelous array of natural habitats across Oklahoma. The dry, short grass prairie and the harsh topography of the Black Mesa region in the Panhandle, the rolling prairies of Osage County, and the more humid eastern forest-covered Ozark and Ouachita mountains all contribute to the geographical fabric of Oklahoma.

These habitats are drawn together in hierarchical drainage basins, but the differences are obvious just as the clear waters of the Glover River bear little resemblance to the broad, flat sandy riverbeds of the Washita or Canadian rivers. These wide habitat patterns are accentuated by the many unique habitats such as the Great Salt Plains, the Little Sahara sand dunes, and the bald cypress bottomlands of the Little River in the southeastern corner of the state.

The paired habitat photographs provided by Milby and Johnson convey the geographical and habitat diversity across our state. Some of the images show dramatic changes between the "before" and "after" photographs. Among the most stark alterations are the conversion of prairies, such as those in Kay and Grant counties, into vast landscapes of winter wheat, punctuated with those farmsteads that have survived consolidation into large farming operations. Similarly, cutting timber from the oak-pine-hickory forest in the eastern regions resulted in immediate, wholesale changes, some of which persist today.

The fate of the forests took several pathways, some slowly recovered from timber and eventually replaced the original species composition, others were converted into pine plantations, and still others became farmlands, pastures, towns, and cities. Overall, native vegetation has been altered or eliminated by many human activities, including urbanization, installation of transportation systems, harvesting of natural products, agriculture, grazing, invasion by non-native species, changes in stream flow patterns, and other human-driven influences.

On the other hand, in many of the paired photographs, change over the past 50 to 100 years has been very minimal. Indeed, the authors describe differences over 50 years or more in terms of the invasion by a few more red cedars or only modest increase in the number of rangeland shrubs. As depicted in the pictures, the vegetation is quite resilient, either resisting human use and influences, or by recovering. Thus, this photographic journal lets us both appreciate the tenacity of the vegetation in some instances while simultaneously lamenting the loss of biological diversity in other comparisons.

Early Oklahoma explorers saw our state through a different lens from ours today. They noted the grandeur and beauty, but also saw the settlement opportunities in the vastness of the "unassigned land." Building railroads, driving cattle across the plains, plowing the prairie sod for crops, cutting timber for building and export, mining coal and lead, and damming rivers and streams for water supply and flood control were the work of progress. The photographs of the great land runs between 1889 and 1903 largely convey the notion of human opportunity, as do the scenes of railroad construction. Similarly, the cowboys who drove thousands of cattle along the cattle trails from south to north in western Oklahoma were romanticized as folk heroes.

These measures of progress, however, conflicted with the Native American populations, both those already in Oklahoma and those relocated here. Cattle herds, land runs, railroads, barbed wire fences, coal mines, and settlements were fundamental intrusions into native lands and into the culture of the peoples who were native to our state. The images in this book make these actions, and their consequences, very clear. The admiration for the early settlers who endured so much to carve out a living is surely tempered, even overwhelmed, by the discomfort of knowing that cultures and ways of life were changed or lost forever. Natural ecosystems, such as those that were home to Native Americans and those that exist today, provide an array of values. These values can be divided into two large classes, those that are used by humans and those that are not but represent an intrinsic value just by their existence. The uses can be direct, such as harvesting timber in McCurtain County; non-consumptive, such as recreation in Dripping Springs State Park. Indirect, or values outside the ecosystem itself, are upstream forests that capture nutrients that would otherwise reduce water quality in Grand Lake. Retaining the availability for future use represents option value. The photographs in this book depict these uses in ways far more poignantly than text alone.

Oklahoma Landscapes can be engaged at many levels. At the most general level, one can appreciate the breadth and diversity of our landscapes. These observations can be refined at a second level by comparing the paired images, noting that in some cases there are dramatic differences over time, but in others, the vegetation has changed very little. But, there is another layer made conspicuous by the book, namely those of competing or conflicting uses of our natural resources spanning nearly 200 years. This pictorial journey exposes these layers of understanding with great clarity.

Our understanding of the vegetation changes across Oklahoma is made possible by previous and ongoing research, including that of co-authors Johnson and Milby and Shutler and Hoagland. We owe them all a debt of gratitude because by their work we can far better understand ourselves and our state.

INTRODUCTION & ACKNOWLEDGMENTS

Oklahoma has entered into its second century as the 46th state of the Union. The march toward statehood began on April 22, 1889, in the most famous and unique land run in world history. That opening, and those that followed, resulted within a few years in the settlement of both Indian and Oklahoma territories, and changed forever the landscapes of the future state.

Those changes to Oklahoma's landscapes are both persistent and profound and have led to the development of a thriving population and prosperous economy within the state. At the same time there is an abiding interest in Oklahoma as it was before these changes occurred. How have Oklahoma's prairies and plains, mesas and mountains, and rivers and streams changed as a result of settlement? How did Oklahoma landscapes appear to its original inhabitants and to the early explorers before farms and cities came upon the scene?

Much of the appearance of those original landscapes is captured in the writings of early travelers in the Oklahoma. Thomas Nuttall and Dr. Edwin James are two of the better known explorers of the territory whose diaries and official reports include descriptions of plant life and vegetation of the country through which they passed. Other travelers who left valuable accounts of the land that became Oklahoma are Josiah Gregg, a merchant in the Santa Fe trade who traveled through the area several times in the 1830s and 1840s, the noted author Washington Irving, who made a tour of the prairies in 1835, Captain Randolph Marcy, who led two United States Army expeditions to explore the territory in the 1840s and 1850s, and Dr. Samuel Woodhouse, the surgeon-naturalist for the Creek Boundary Survey in 1849 and 1850.

The records of these early nineteenth century travelers give us an impression of a pristine land inhabited only by wildlife and native people. While these written reports are irreplaceable as information sources about pre-settlement Oklahoma, they nevertheless give way to the more exact and accurate impression that can only be achieved through photography.

A legacy of pictures left to us by pioneer photographers is of great interest for the record which they provide of landscapes in early Oklahoma. The first known photographs in what is now Oklahoma were several daguerrotype made in 1850 by W. C. Mayhew, a member of the Creek Boundary Survey party. Unfortunately, Mayhew's work did not include any landscapes which could later be identified.

The earliest landscape photographs of identifiable locations that we found are those of William Soule who was in the Fort Sill area from 1869 to 1875. At least one of his early photographs includes an image of his own darkroom tent which was erected on the banks of Medicine Creek near Fort Sill. Its presence in the photograph reminds the viewer of the hardiness and dedication of this and other early photographers and evokes appreciation for their efforts to preserve for us the scenery of the nineteenth century world.

A number of early photographers followed Soule in documenting landscapes of Oklahoma on film. Perhaps the best known were J. Willis Stovall and Charles N. Gould, scientists at the University of Oklahoma. Stovall's pictures of sites in the Oklahoma Panhandle and Gould's photographs taken in the Arbuckle Mountains provide records of some of the most distinctive landscapes of the state.

The work that follows includes photographs made by pioneer photographers with a matching photograph taken a half century or more later documenting changes that have occurred in the landscape.

In addition to our obligation to these early photographers, we are indebted to a number of other persons for their assistance and support during the conduct of this work. John Lovett, photo-archivist and librarian of the Western History Collection at the University of Oklahoma, provided access to the photographic archives at the University Library. Towana Spivey, curator of the Fort Sill Museum, was helpful and generous in providing access to the collection at the museum and in making copies of a number of important photographs from the museum collection.

Claudine Daniels, Environmental Education Officer at the Wichita Mountain Wildlife Refuge, was helpful in making copies of pictures for our use from the refuge archives as well as giving us directions for relocating sites of interest within the refuge. Chuck Milner, Range Land Management Specialist, provided photographs of Black Kettle National Grassland.

In addition to these archivists we are especially grateful to Drs. Gary Schnell and Caryn Vaughn for their generous financial support for the field and photographic work during their tenures as director of the Oklahoma Biological Survey . We owe an immeasurable debt to Gregory Strout of the Samuel Roberts Noble Electron Microscopy Laboratory at the University of Oklahoma for enhancing archival photographs and restoring them to publishable quality.

We are indebted to Daniel Hough of the Oklahoma Biological Survey for his guidance through the process of photo reproduction, to Dr. Bruce Hoagland and Todd Fagin of the Oklahoma Biological Survey for map production, as well as technical and editorial advice during the production of the manuscript and Ed Shaw for reading the manuscript and offering many suggestions for its improvement. Finally, we are grateful to Dr. Gordon Uno of the Department of Botany and Microbiology for his suggestions for improvement of the pictures and text.

The successful completion of this project could not have been accomplished without the help of each of these persons. At the same time, the authors take full responsibility for the accuracy of the content of the work.

T. H. Milby
Forrest L. Johnson

OKLAHOMA BEFORE STATEHOOD

When Oklahoma was opened for settlement toward the end of the nineteenth century, it was a land largely unchanged by the influx of European settlers who had significant impact on the surrounding territories. The only visible influence came from the forced relocation of native peoples from the southeastern region of the United States who had brought crops, cattle ranching, and other commercial activities to Indian Territory as the tribes continued to practice enterprises developed through contact with European settlers in their former home.

While the relocation of eastern tribes in the 1830s and 1840s brought new influences, these were not the first people to leave impressions on the Oklahoma landscapes. Many of the earliest inhabitants were wandering hunters and food gatherers whose traces are found largely in caves and cliff overhangs which they occupied either continuously or intermittently as they forayed across the countryside to hunt game and collect plants for food. Their impact on the landscape was small, and has been largely obliterated by the continuing forces of nature.

Those wandering and nomadic people were followed by the mound builders who lived in settlements near the Arkansas River in eastern Oklahoma beginning about 500 A.D. Numerous burial and ceremonial mounds, some nearly fifty feet high, remain as testimony to these peoples' engineering skills and are evidence of their impact on the landscape. By the time of the arrival of the earliest European explorers to Oklahoma, the Spiro people had been replaced along the Arkansas and Red rivers by people of the Caddo, Wichita, and Quapaw tribes. These were hunters and farmers living in permanent villages and, unlike their Spiro predecessors, left little permanent imprint on the landscape.

It was into this largely empty and unmarked land that Indian tribes from other states and territories of the United States were relocated in the early part of the nineteenth century. Although the Cherokees and other members of the civilized tribal groups brought with them an agricultural and trading economy, their

A group of cowboys drop for the noon meal on the Chisholm Trail. Cattle grazing and their movement through the territories to northern markets were major enterprises in the nineteenth century and brought changes to the landscape of Oklahoma prior to its opening for settlement. The Chisholm Trail was one of the most famous and heavily used of the numerous trails. Historical markers at several points identify the route of the Chisholm Trail where it crosses highways in the central part of the state today. The two children at left may be visitors from a nearby ranch. *Courtesy Phillips Collection, University of Oklahoma Libraries.*

populations were small and did not significantly change the vegetation of the new region to which they came.

While opening of the territories of Oklahoma to European settlement in the latter part of the nineteenth century had a larger influence on natural vegetation, a number of lesser, but nevertheless significant, cultural enterprises did leave their imprints on the landscapes of the state. Among the most significant and lasting were the movement of cattle across the territories from Texas to markets in the north, migration trails of settlers and traders from east to west both before and after the Civil War, the construction of railroads, operation of coal mines, and the harvesting of timber mostly in the southeastern part of the state.

The movement of cattle from the Texas range to northern markets began as early as the 1840s. Fifty thousand head of cattle crossed the Red River into Oklahoma in a single year bound for Kansas City and other towns which provided access to eastern railroads. The Civil War interrupted this movement through the territory only to see the flow of cattle resume following the conflict and increase to a number in excess of 500,000 during a single year. By some estimates six million cattle moved through Oklahoma in the two decades following the war.

The movement of these vast herds of cattle followed several well-established trails. The Shawnee Trail consisted of two branches leading from Boggy Depot in southeastern Oklahoma across the eastern half of Indian Territory to outlets in Kansas and Missouri. Probably the most famous trail, named for Jesse Chisholm, was used to move cattle from Texas to Kansas approximately parallel to the 98th Meridian.

Another trail, the Great Western, passed through what later became Greer County into the Kiowa-Comanche reservation across Cheyenne-Arapaho country, the Cherokee Outlet, and to the cattle pens at Dodge City, Kansas. Movement of large herds had a significant influence on the grassland over which they passed. Cattle were moved slowly through the territory, often covering 10 to 15 miles per day, with occasional layovers allowing cattle to fatten on the native grasses.

In some cases, large herds were wintered on the native pastures before resuming the northward march with the arrival of spring. Movement of large herds aroused the enmity of the Indian tribes through whose reservations the cattle passed. The Indians did not like the unauthorized intrusion of the herds on their land and the disturbance caused to buffalo and other game on which the Indians depended for their living. While little evidence remains today to mark the existence of the trails, their use signaled the eventual destruction of the buffalo herds and of their ultimate replacement by domestic cattle.

While the movement of large numbers of cattle across the territory had temporary impact on Oklahoma vegetation, other activities that took place during the middle decades of the

By the late 1880s, buffalo had disappeared from the prairies of the West, replaced by cattle whose numbers increased in the decades of the open range after the Civil War. Following settlement, fences restricted grazing to limited areas, often resulting in overgrazing and changing the range composition from palatable and nutritious native grasses to weedy grasses and forbs. In many cases native pastures were broken for cultivation or planted to non-native forage plants. In either case the disappearance of buffalo was followed by major changes in the vegetation of most of the prairie land of Oklahoma and the central plains. *Courtesy Forbes Collection, University of Oklahoma Libraries.*

nineteenth century were less benign. Of more lasting impression on the landscape was the movement of emigrants and commerce along several wagon and stagecoach trails during the mid-part of the century.

One of the first to come into common use was the Santa Fe Trail. The Cimarron Cutoff, a south branch of the trail, passed through what is now Cimarron County at the western end of the Oklahoma Panhandle. This shorter southern route provided passage around the difficult terrain of the original northern route through Raton Pass, while at the same time exposing travelers to the privations of the waterless high plains and hostile action from Indian tribes through which the cutoff passed.

Originally used to accommodate trade between merchants in Missouri and Santa Fe, it later served as a route for emigrants to New Mexico and California. It is estimated that by 1860, as many as 6,200 mules and oxen, 3,000 wagons, and 7,000 men had been engaged in trade and travel to New Mexico on the Santa Fe Trail. A train of several score of wagons in a single party might move in a fan, 100 yards wide, their tracks resulting in the formation of a depression as wide as the phalanx of wagons themselves.

The tread of thousands of ox, mule, and horse hooves compacted and denuded the soil. Subsequent blowing and washing resulted in a permanent depression on the plain. In the case of the Santa Fe Trail, this depression caused by passage of immigrant wagons to the west can still be seen in Cimarron County where it stretches to the horizon on the boundary between Oklahoma and New Mexico.

A number of additional trails provided passage for immigrants and commerce through the territory. The California Road ran from east to west beginning at Ft. Smith, Arkansas, followed the Canadian River valley to the region of prominent buttes in the west to the Texas panhandle. Two additional trails, the Butterfield Stage Route, established in 1858, and the Texas Road brought travelers respectively from Ft. Smith and from Kansas and Missouri at the northeast boundary of Indian Territory to the Red River and to Texas. A number of trading centers along these trails including Salina, Fort Gibson, and Atoka, are present-day towns; many others have vanished.

One of the most heavily traveled trails in Oklahoma was the Cimarron Cutoff of the Santa Fe Trail which was a major trade route between St. Louis, Missouri, and New Mexico. Ten thousand wagons are believed to have made the trip across the prairie on the south branch of the Santa Fe which crossed the western part of the panhandle in what is now Cimarron County. Ruts and caravan impressions are preserved in portions of the trail well over a century after it fell into disuse and can be seen in this photograph made by an unknown photographer in 1938. The shadow cast across the lower left side of the picture may have been caused by the unknown photographer's vehicle parked nearby. *Courtesy Western History Collection, University of Oklahoma Libraries.*

Before the last decade of the nineteenth century, a number of major rail lines had crossed Indian and Oklahoma territories, each imposing major and lasting imprints on the landscape. By the end of the Civil War, each Indian tribe in the territories was required to allow construction of railroads through its reservation. The earliest of lines to come into the territory was the Missouri, Kansas, and Texas, also known as The Katy, which passed through the Cherokee Nation beginning at the Kansas border near Chetopa and extended as far south as Chouteau by 1871. Following closely the route of the Texas Road, the M K &T ran southwesterly into the Choctaw Nation and to the Red River near Sherman, Texas.

Second only to railroads in impact on the landscape of the territory was coal mining and the use of coal. Although coal was mined in the Choctaw Nation prior to the Civil War, its use primarily was for fuel in blacksmithing and other local needs and resulted in relatively minor impact on the countryside. Development of distant markets in the mid-1860s brought a more substantial coal mining industry to Oklahoma.

Coal deposits in the Choctaw Nation were the first to be exploited by commercial mining. Mining continued to expand into coal deposits in what later became Tulsa, Rogers, Wagoner, and Craig counties, all part of the Cherokee Nation. By 1900, more than two million tons of

Railroads came to Oklahoma with the relocation of native peoples to Indian Territory in the early part of the nineteenth century. Treaties with the tribes provided for railroad right-of-ways across all of the major tribal lands. An early photograph taken sometime in the 1890s is of workmen laying railroad line directly on the level prairie. *Courtesy Forbes Collection, University of Oklahoma Libraries.*

Railroad construction ran through rough country near Concho Indian School in Canadian County. Over uneven country, the railroad line required drastic changes and cuts in order to meet grade requirements. Railroads have been a major factor in the alteration of Oklahoma landscapes in the state's first century of development. Nearly 4,000 miles of rail lines were in active operation in 2007. *Courtesy Shuck Collection, University of Oklahoma Libraries.*

coal was being produced annually from mines in Indian Territory.

While the original mining was an underground process, strip mining soon resulted in the formation of mounds of tailings, which in most cases became a permanent part of the landscape of the mined area. More recently strip-mined areas have been restored to contour as the result of the Surface Mining Control and Reclamation Act of 1977. Most of the earlier mine sites remain un-reclaimed. Where soil conditions support its establishment, vegetation has returned to mounds which lie beside the abandoned pits. Many of the older mounds have taken on the appearance of the surrounding terrain and belie the earlier disturbance of the landscape by strip-mining operations. Although the older abandoned mine areas appear to be less intrusive, their presence remains a permanent imprint on the landscape much as embankments

A steam shovel between two mounds of overburden it has created to expose a shallow vein of coal. *Courtesy Herring Collection, University of Oklahoma Libraries.*

Coal mining became an increasingly important industry in Indian Territory in the decades following the Civil War. More than two million tons were produced annually by the turn of the century. *Courtesy Oklahoma Geological Collection, University of Oklahoma Libraries.*

An early-day Oklahoma sawmill in the piney woods of southeastern Oklahoma changed forever the original vegetation of the region. *Courtesy Oklahoma Geological Survey Collection, University of Oklahoma Libraries.*

The future site of Oklahoma City on April 22, 1889. The freight train in the background waits on the Atchison, Topeka, and Santa Fe railroad tracks. *Courtesy Forbes Collection, University of Oklahoma Libraries.*

The population of Oklahoma City had grown to 12,000 within a few days of the 1889 Land Run. The addition of buildings and streets brought great change to the prairie landscape of Oklahoma. *Courtesy Forbes Collection, University of Oklahoma Libraries.*

and cuts are permanent evidence of the existence of railroads on the terrain through which they pass.

Along with development of the coal-mining industry, the arrival of railroads into the territories led to commercial exploitation of the forests for lumber and other wood products, especially in the oak-hickory-pine forests of the Ouachita Mountains of southeastern Oklahoma. Prior to construction of the Missouri Kansas & Texas Railroad into the Cherokee, Creek and Choctaw nations in 1872, timber harvesting had been confined largely to the Arkansas River and its larger tributaries. Logs were floated downstream to Fort Smith, Arkansas, and other potential market towns.

The coming of railroads into the mountain regions of the southeast brought access to extensive pine, bald cypress, and hardwood forests of the region. Sawmills and boomtowns sprang up along the railroads and thrived until all merchantable trees had been harvested from adjacent slopes and valleys. In some instances, portable saw mills were assembled in more remote areas of the virgin forest from which logs were transported by mule teams or oxen.

Logging activity continued in ever more remote areas until virtually all of the original forest was gone. At that time, many of the towns whose lives and economies were dependent on the harvest of old-growth forests died.

One of the last sawmill towns was Pine Valley, which survived until 1942 when all of the useable timber had been cut. A decade later the railroad and all buildings of the town were dismantled and removed. The town, along with the virgin forests on which the life of the town depended, was gone.

The effect of the original "cut out and get out" practice of forestry in Oklahoma and throughout the South changed forever the vegetation and landscape of the region. In most cases the land on which the original forests had stood was abandoned and allowed to return to a vegetation type that resembled, at least in species composition, though not in size, that which characterized the area prior to logging. However, the quality of the forest was in most cases severely degraded.

Scorched earth and a plume of smoke in the background suggest that fire was used to clear the land of all but the largest trees prior to the original cultivation in the wooded parts of the state. "Blackjack country," as this early-day photograph is named, probably refers to the Cross Timbers which is the post oak-blackjack forest lying in a belt north to south across the mid-section of Oklahoma, and extending into Texas and Kansas. *Courtesy Wenner Collection, University of Oklahoma Libraries.*

Deep, rich soils that supported the native tall grasses of the prairie are ideal for the growing of winter wheat. Much of the original tall grass prairie has been converted to cultivation and only remnants of virgin prairie remain. Usually these samples can be found over rocky sub-soils as in the Tall Grass Prairie Preserve in Osage County or where early intervention established their protection as in the case of the Wichita Mountains Wildlife Refuge. In 1898 a 13-horse plow breaks the prairie somewhere in northern Oklahoma. *Courtesy Phillips Collection, University of Oklahoma Libraries.*

The magnitude of the changes to come after the Cherokee Outlet land run is seen in the gathering of horses, wagons, and riders on the south border of the outlet on the day of the opening. *Courtesy Phillips Collection, University of Oklahoma Libraries.*

In recent years much of the area where the original oak-hickory-pine forests stood have been acquired by large lumber and paper product companies. Their practice of clear cutting and replanting to monocultures of genetically improved southern yellow pines has further changed the vegetation of the region from its original character and appearance.

While the intrusion of railroads, mining, and logging made significant and lasting imprints on the landscape, by far the most profound changes to the land were the result of opening land of the two territories which were later to be combined to form the State of Oklahoma.

The first of these openings for settlement was into the Unassigned Lands, 1.8 million acres in the central part of the territories. Into these lands on April 22, 1889, poured an avalanche of settlers eager to claim either a 160-acre homestead or a town lot within one of the several town sites that had been established. An estimated 50,000 people made the run. By November, the Secretary of the Interior reported that 60,000 persons were in the Oklahoma District, the new name of the Unassigned Lands.

The Cherokee Outlet land run on September 16, 1893, was the largest and most spectacular of the several land rushes which brought settlers to the territories of Oklahoma. An estimated 100,000 people took part in the race to claim 30,000 parcels of land in the Cherokee Strip, as it was frequently called. Breaking the prairie for wheat became the major enterprise for most of these original settlers, the accomplishment of which imposed major changes on the landscapes of the state.

Similar openings within the Oklahoma and Indian territories took place throughout the remainder of the decade until opening of the final parcel in southwestern Oklahoma in 1901 concluded the settlement of the state. Population had increased to 390,000 by 1900, the major portion of which consisted of farmers who had come to convert the rich prairies into a land of farms. By 1902, almost the entire state had been settled and all lands suitable for cultivation had been converted to agriculture.

By the 1910 census, the first census following statehood, the population of Oklahoma had reached 1.6 million and continued to increase into the third decade of the century when the severe drought of the 1930s reversed the trend. By 1930, the population of Oklahoma peaked at 2,396,000, before beginning its decline.

Many farms in the northern and western parts of the state were abandoned in great number as families moved from farms to the towns within the state or to California and other west coast states to reestablish their lives in a region which at the time offered greater economic opportunity.

The abandonment of farmland as a result of dry years was one factor of significant change in Oklahoma landscapes. The drought of the 1930s taught land owners that much of the region was better suited for maintenance of permanent vegetation than for cultivation. Farmers and ranchers, with the cooperation and assistance of the United States Department of Agriculture Soil Conservation Service, restored much of the land to original native grasses. Livestock grazing and hay production, rather than cultivated crops, became a major part of the agricultural economy in the drier parts of the state.

In some areas in western Oklahoma, farms were abandoned altogether. These abandoned lands were re-vegetated in most cases to native short grass species and became a part of the Black Kettle National grasslands where they are managed in a natural condition under controlled grazing. In eastern and southeastern parts of the state a similar change in the status of farmland occurred. Abandoned farmland reverted to ownership by the federal government and was incorporated into the Ouachita National Forest where it was replanted to pine trees or allowed to return spontaneously to forestlands similar to those it had supported prior to settlement.

The abandonment of farmland in both eastern and western parts of the state and the reestablishment of permanent vegetation have been major factors impacting landscapes throughout the state since the dust bowl days of the 1930s. Abandonment of cropland and suppression of fire are significant factors in the determination of the character of Oklahoma landscapes. Control of fire throughout the grassland areas of central and western Oklahoma has resulted in a dramatic increase in the establishment and growth of eastern red cedar trees.

Much of Oklahoma, which was grassland at the time of the first opening in 1889 and converted to cultivation a decade later, has now reverted to its original grassland condition or has become red cedar woodlands as a result of the suppression of fire. Natural forces, particularly drought, have reestablished themselves as dominant factors in the establishment of landscape. The development of red cedar forest land within the state is then an appropriate reminder of the blending of both human and natural forces on the creation of the landscapes of Oklahoma since opening the land to settlement.

EARLY PLANT EXPLORERS

Oklahoma is a state of rich and varied landscapes. The state lies near the midsection of the United States and, including its slender Panhandle, extends nearly 500 miles between its eastern and western borders. The wide range of elevations, soil types, and climatic conditions support a number of vegetation types ranging from pinyon-juniper, on the nearly 5,000-foot crest of the Black Mesa in the Panhandle, to the shortleaf pine and deciduous forests at the lower elevations in the southeastern corner of the state.

Precipitation amounts vary from 16 inches in the west to as much as 50 inches along the eastern border. Frost-free days increase from less than 200 days on the high plains to more than 240 days in the woodland regions of the east. Each of these factors contributes to the vegetational diversity of the state.

Shortgrass species such as buffalo and grama grasses are the dominant species of the high plains of the Panhandle. Sand sage, oaks, and various shrubs grow on sandy soils and stabilized dunes along rivers in the northern part of the state. Similar sandy soils support shinnery oaks and sand sage along the western border. Blackjack-post oak forests and tall grass prairies cover a large part of the central region of the state, although most of these prairies were converted to agriculture following settlement. Oak forests have returned to much of this area where farming has been abandoned.

In his painting, Wayne Cooper preserves Washington Irving's meeting the Osage on the banks of the Arkansas River near present-day Tulsa during Irving's tour of the prairies in 1832. *Courtesy Oklahoma State Senate Historical Preservation Fund, Inc.*

Oaks, hickories, and shortleaf pines dominate the Ouachita, Boston, and Ozark mountains in the east, while post oak, blackjack oak, and eastern redcedar are characteristic of the Arbuckle and Wichita mountains in the south and southwestern parts of the state. Bottomland forests throughout the state commonly include black willow, cottonwood, American elm, ash, hackberry, and sycamore.

Although several types of vegetation have wide distribution in general, the number of tree species decreases and the size of both trees and grasses diminishes from east to west across the state. While post oak-blackjack oak forest and tall grass prairie are the predominant natural vegetation types, there also are representatives of Rocky Mountain foothill vegetation, short grass prairies, Ozark hardwood forests, and forests of the Gulf Coastal Plain.

This diversity of vegetation has been of immense interest to botanists and geographers. Numerous nineteenth century travelers and explorers provided written descriptions of the plant life, geology, and fauna of the territory that was to become the State of Oklahoma. In 1819, Thomas Nuttall was the first to leave extensive records of his travel into the region. He traveled south along the Poteau and Kiamichi rivers into Choctaw territory "...scattered with oaks and pines...," noting, "...the whole country a prairie, full of luxuriant grass about knee high..." interspersed with "...somber belts of timber, which serve to mark the course of rivulets." Later journeys took Nuttall north along the Grand River westward through the central part of the territory, then back to Ft. Smith along the Cimarron and Arkansas rivers. His journal includes daily reports of the plant life and topography of the region.

Shortly after Nuttall's exploration of the eastern rivers of the region, Major Stephen H. Long followed the Canadian and Arkansas rivers through the territory in 1820 returning from his expedition to the Rocky Mountains. Reports by Edwin James, botanist for the expedition, include observations on particular plants along the way as well as descriptions of the forests and prairies which covered the bottomland.

Typical of these descriptions is that of the vegetation observed at the confluence of the Canadian and Arkansas rivers in September, 1820. "Thick and extensive cane brakes occurred on all sides, and though the bottoms were wide and covered with heavy forests we could see, at intervals, the distant sandstone hills with their scattered forests of cedar and oak." John Torrey later published a partial list of the plants collected by James.

The account of Washington Irving of his journey through what is now the eastern part of Oklahoma as far south as the Canadian River includes frequent but cursory references to prairies and forests in 1832. Typical of his descriptions is that of the forests on the Arkansas River at their crossing point a short distance above its confluence with the Cimarron, then called the Red Fork. "The river scenery at this place was beautifully diversified... bordered by willows and cottonwood trees, rich bottoms, with lofty forests; among which towered enormous plane trees, and the distance was closed in by high embowered promontories."

At his crossing of the Cimarron, some 75 miles north of its mouth, he encounters the Cross Timbers which:

> *...is about forty miles in breadth, and stretches over a rough country of rolling hills, covered with scattered tracts of post oak and black-jack; with some intervening valleys, which, at proper seasons, would afford good pasturage. It is very much cut up by deep ravines, which, in the rainy season, are the beds of temporary streams, tributary to the main rivers, and these are called branches.*
>
> *The whole tract may present a pleasant aspect in the fresh time of the year, when the ground is covered with herbage; when the trees are in their green leaf, and the glens are enlivened by running streams. Unfortunately, we entered it too late in the season. The herbage was parched; the foliage of the scrubby forests was withered; the whole woodland prospect, as far as the eye could reach, had a brown and arid hue.*
>
> *The fires made on the prairies by the Indian hunters, had frequently penetrated these forest, sweeping in light transient flames along the dry grass, scorching and calcining the lower twigs and branches of the trees, and leaving them black and hard, so as to tear the flesh of man and horse that had to scramble through them. I shall not easily forget the mortal toil, and the vexations of flesh and spirit, that we underwent occasionally, in our wanderings through the Cross Timbers.*
>
> *It was like struggling through forests of cast iron.*

This description of vegetation as well as of wildlife and of the activity of the native people provides a picture of the region before changes brought about by settlement.

In 1834, Lieutenant T. R. Wheelock chronicled events along the route of a regiment of United States Army Dragoons under the command of Colonel Henry Dodge. Beginning at Fort Gibson, the regiment explored the southern part of the territory to the site of old Camp Washita near the Red River and west to the Comanche villages on the North Fork of the Red. Wheelock notes passage of the troops through prairies, Cross Timbers, and scrub oak at many stations. On August 8, near Little River, he wrote, "The timber is larger here. Black walnut and sycamore; lime and freestone. The woods abound to-day in plums, and a variety of finely flavored grapes...." The map of his route provides additional records of the vegetation at specific dates and places along the line of travel.

Another nineteenth century traveler in the territory was Captain Nathan Boone, son of pioneer explorer Daniel Boone. In 1843, Boone led a regiment of dragoons northwest from Ft. Smith along the Salt Fork of the Arkansas through the Osage Indian reservation and north

Nathan Boone, son of Daniel Boone, made several expeditions through what would become Oklahoma and recorded his travels in reports and journals. *Courtesy Oklahoma Historical Society.*

to central Kansas. Returning from Kansas he joined the Cimarron at the confluence of the North and South forks of that river.

On the 67th day of his journey, near the divide between Little River and the Canadian, he reported, "Skirted the cross timbers for about two miles. The timber is so-called, from the fact it runs from the Red River across to the Arkansas.... It consists of dwarfish scrub oaks, whose branches extending down and interlacing render it almost impenetrable." The journal gives repeated references to vegetation and plant life found along his return route to Ft. Smith.

In 1845, Lieutenant J. W. Abert led an exploratory expedition from Bent's Fort in Colorado to St. Louis, Missouri. Exploring the upper reaches of the Arkansas, Canadian, and Washita, his party entered the territory of Oklahoma at the 100th Meridian and followed the course of the Canadian across the body of the territory to its confluence with the Arkansas. His journal included observations on the flora and fauna of the region, often providing exact readings of latitude and longitude at which his observations were made.

On September 23, between the Washita and the Canadian, his journal reported "... rough and broken... country. The bottoms of the streams were very well timbered; and among the various trees we noticed, the cotton-wood, elm, black locust, 'Robinia pseudo acacia', black walnut, 'Juglans nigra', and the coffee-nut tree, 'Gymnocladus Canadensis', which contains, in its large legumes, a very palatable nut, which we collected and ate."

A few days later, at the 97th Meridian, Abert wrote, "We now traveled on a narrow strip of land between the Canadian and its north fork, which appears to be the more heavily timbered of the two; for in looking in the direction of the north fork, we saw a dense forest of oaks extending as far as the eye could reach."

In 1850 and 1852, Captain R. B. Marcy conducted expeditions into regions of the territory providing descriptions of the wildlife, vegetation, and topography. The first of these charted the route of a proposed railroad from Ft. Smith to Santa Fe, New Mexico. His course followed the Canadian River to the region of the Texas high plains. In addition to notes on vegetation at stations along the route, he provided general observations on the wooded and mountainous nature of the eastern part of the region, "Beyond which we struck plains ... yet there is no place upon the whole road where a very long drive has to be made to get water."

Marcy's second expedition two years later provides the first description of the course of the Red River, including its two main northern tributaries. On the North Fork of the Red he observed, "The only varieties of timber that we find upon this part of the Red River are cotton-wood and hackberry, the former greatly predominating and of large dimensions. Indeed, I have never seen so much timber at any other place upon the plains, in this longitude, as we find here."

In 1853, Lieutenant A. W. Whipple led an expedition west from Ft. Smith following a route along the Canadian River similar to the one taken by Marcy three years earlier. Whipple's journal provides notes about vegetation and plant life seen along the way. Locations of his campsites are noted on his chart to which observation of plant life can be related. At Camp 34 on September 3, he reports "...leaving 'Oak Creek' we crossed a rolling prairie six miles to Buffalo Creek. This is a well timbered ravine with large alamos & oaks.... A species of grama called buffalo grass being now predominant & buffalo signs covering the prairie suggest the name given this creek."

These explorers and travelers through the territory of Oklahoma in the early half of the nineteenth century provide us with a picture of landscapes and vegetation prior to its opening for settlement. And while it is possible to make assessments of the changes that have taken place in Oklahoma vegetation and landscapes based on written descriptions of these early explorers, a far more accurate and precise analysis of these changes is possible only when they are based on photographic records of the region.

The advent of the camera, especially as it became sufficiently portable to be operational outside of the studio, made possible the recording of the landscape with a precision and objectivity not previously possible with the written word alone.

The Oklahoma frontier became an attraction for a number of photographers with an eye for creating a camera record of people and landscapes well before the opening of the territory to settlement. These early records, many of which are preserved in archives and museums, provide a compelling opportunity to examine and document photographically changes that have occurred in the vegetation and to the landscapes of the state in the century or more since the influx of settlers in the late 1880s. It is this kind of comparison based on older and more recent photographs of which the present study consists.

The earliest of the pioneering photographers was William S. Soule, whose photographs taken before 1875 provide a record of landscapes around Fort Sill while it remained largely unchanged by human activity. Other photographers whose work capture early landscapes within the state include Charles N. Gould, the pioneering geologist, and J. Willis Stovall, early paleontologist, both of the University of Oklahoma.

Fortunately, many early photographs,

A drawing represents Captain Randolph B. Marcy's camp in Beckham County, June 6, 1852. *Courtesy Oklahoma Historical Society.*

although unidentified as to date and to source, provide sufficient description of place of origin that the site location can be identified today. The present study relies on these early eyes as a basis on which comparative late twentieth century photographic studies can be made. Using these early photographs, it is now possible to record and describe changes that have occurred during the century since Oklahoma was first opened for settlement.

The photographic journey that follows includes snapshots of early day landscapes of the Sooner State. Each of these early photographs is matched by a more recent picture which reveals the extent to which these landscapes have been influenced and changed by settlement since the great land rushes.

Changes that have occurred during this first century of development are both indelible and profound. Cities and towns dot the land where virgin prairies and forests formerly stood. Roads and highways crisscross the state. Impoundments and dams have changed the flow of all major rivers and many of the smaller streams. Forest lands have been cleared for farms and the prairies have been broken for wheat and other cultivated crops.

Population has grown from a few thousand scattered native people to more than three million, most of whom live in large urban areas and smaller towns. While changes have occurred throughout much of the state, numerous natural landscapes remain. Many of these less disturbed places are in more sparsely populated regions of the west and in the mountainous and wooded areas of the east and south. A number of these natural places are included in this photographic record and the traveler throughout the state can discover many more.

While these many natural areas provide numerous benefits, perhaps the most generally recognized value of natural areas is the contribution they make to the aesthetic qualities of the state. Oklahoma's natural beauty is increasingly recognized and is promoted as an attraction to both local and out-of-state visitors.

These areas of natural beauty are a source of great economic value. Rugged buttes, grazing buffalo, mystic prairies, and lofty groves of trees provide Oklahoma with a variety of outdoor activities and attractions. And while the aesthetic value of these places is well recognized, areas where human activity is restricted and controlled are of value to the state in other ways.

Natural landscapes and vegetation remind us of the limits of humankind and of our proper place in the scheme of life. They are testimonies to Oklahoma's recognition that the intrinsic worth of the natural world exists independent of utilitarian values nature may also possess. Protection of these virgin forests and prairies declares our recognition that value does not lie in the fulfillment of humankind's material needs. A people's stature is increased by recognizing this value.

In addition to philosophical considerations, and aesthetic and economic values of natural areas, are other practical considerations that justify their preservation. Among these more practical considerations are the genetic variability and diversity which they preserve, especially as these are available for scientific investigation and research. The pool of genetic material contained within organisms in natural areas provides a resource of inestimable value for the improvement through technology of existing plant and animal resources as well as for the discovery and development of additional resources to add to their number.

Of practical consideration also is the increasing awareness of the importance that complexity may have to the stability of ecosystems. Complex systems may be more stable than simple ones. There is the role undisturbed communities play as baselines against which changes brought about by human activity can be measured and by which the health of the environment determined. Their role in the monitoring of environmental health and stability becomes increasingly critical as human numbers and human influences on the environment increase.

This role as standards for the measurement of environmental health becomes more important as the state faces change as it moves into the twenty-first century.

A PHOTOGRAPHIC JOURNEY ACROSS OKLAHOMA

In the following pages of photographs you will find changes that have occurred in the landscapes of Oklahoma since the camera was used to record scenes in the Sooner State. The earliest known photograph was made by William Soule at Fort Sill during the time of its construction in the late 1860s.

A number of photographers, both known and unknown, followed Soule, adding to the archival record of landscapes across the state from near the border with Arkansas on the east to the Black Mesa country in the Panhandle. The most recent of these archival photographs are dated in the mid-1940s. Each of the early photographs gives us an identifiable site where matching more recent pictures were made toward the end of the twentieth century. Seventy-one matched pairs of pictures are included in the collection— the older one showing the site as it appeared 50 or more years ago—and a more recent one showing how the site looks a half a century or more later.

Unfortunately, the photographs are not evenly distributed. This lack of uniformity can be explained by the fact that many of the photographers were pioneer geologists and paleontologists who were drawn to those parts of the state where their scientific interests could be served. The result is that many pictures were taken in a relatively limited number of areas which were found to be of particular scientific interest, while other parts of the state were neglected. Nevertheless, there is no major vegetation type or geographical area of the state which is not represented by one or more pairs of pictures.

Each photograph is identified within one of five geographical regions of the state. The five regions consist of four within the main body of the state in addition to the Panhandle. Photographs are numbered consecutively, beginning at the northeast region and ending at the west edge of the Panhandle.

One who is interested in taking this photographic journey across the state to observe its geologic and vegetational features may locate the site of each photograph by using the map that accompanies each region and the description provided by the caption of each pair of photographs.

The comparative photographs of *Oklahoma Landscapes* are divided into five regions.

NORTHEAST OKLAHOMA

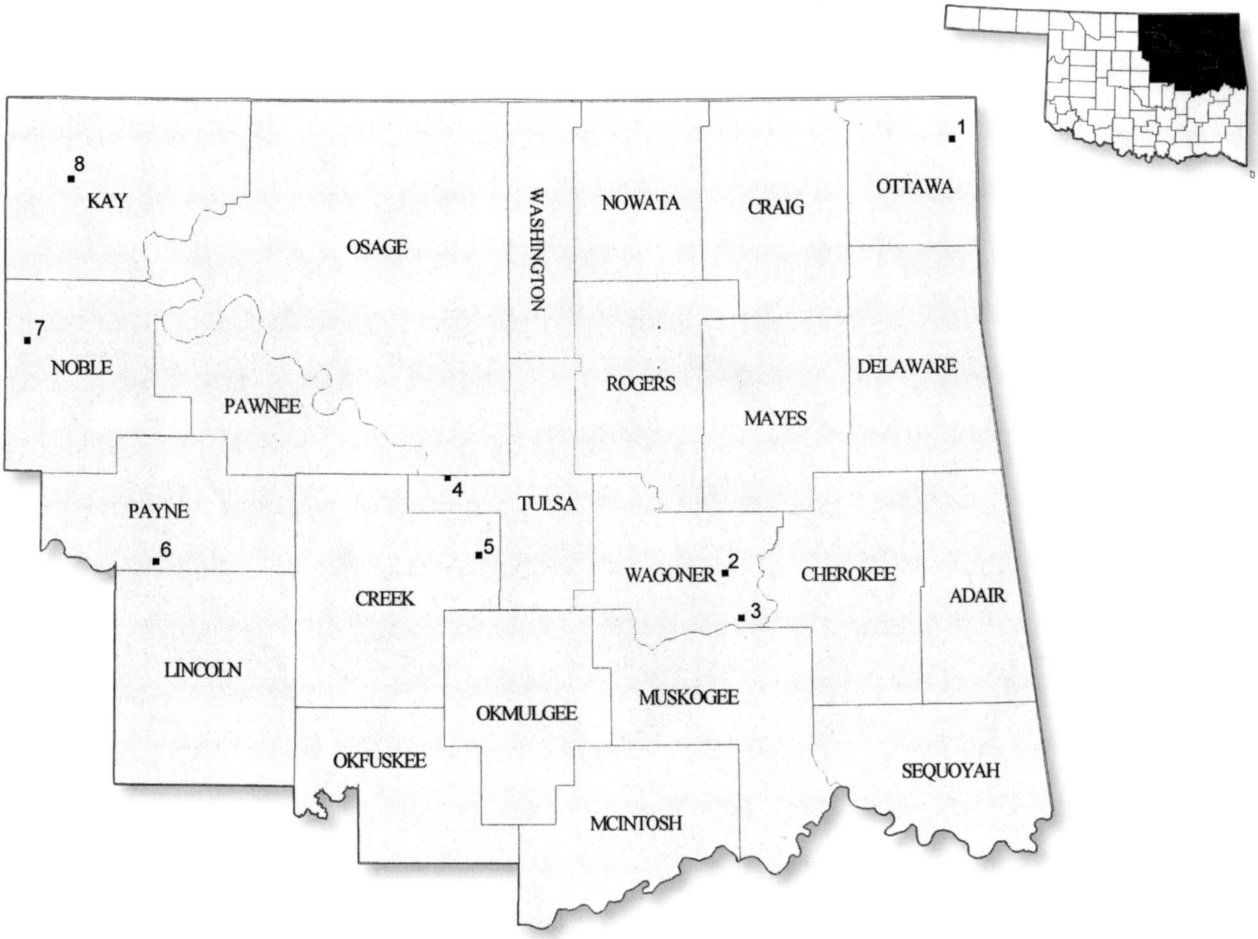

NUMBER ONE

EARLY PHOTOGRAPH

An unusual natural bridge in Ottawa County is the subject of this undated Oklahoma Geological Survey photograph probably taken sometime in the 1920s. The bridge, which allows passage only in a stooped position, is a few yards east of a small cave where a spring flows to a nearby creek. The arch itself may be an isolated span formed by the collapse of a pre-existing cave.

Little vegetation is visible in the older photograph except for a few vines on the rocky soil that surrounds the bridge. The trunks of three large trees can be seen beyond the arch in the background of the picture. In 1989, the immediate vicinity of the bridge had become more heavily vegetated, dimming the light which passed through the arch.

NATURAL BRIDGE NEAR PEORIA

MORE RECENT PHOTOGRAPH

The arch itself was covered with Virginia creeper, poison ivy, and several kinds of herbaceous plants, including pokeweed and knotweed. Trees growing nearby included sycamore, red oak, and black walnut. Both the cave and the bridge are at Cave Springs Ranch, owned in 1989 by Clarence Burr, whose residence was adjacent to the site of the two formations. The site is eight miles east of the Miami exit on the Will Rogers Turnpike and 3 miles north.

NUMBER TWO

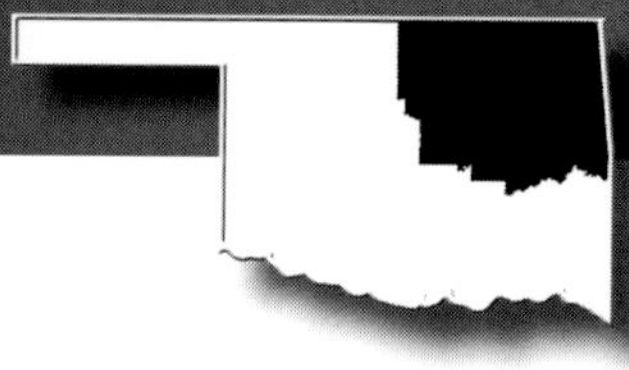

EARLY PHOTOGRAPH

An Indian mound in Wagoner County is the scene in an Oklahoma Geological Survey photograph at the University of Oklahoma Libraries dated May 10, 1928. Two figures on the mound are identified as Professor Spier and Mrs. Searcy.

The location is described as the Norman farm in the Grand River bottoms eight miles east of Wagoner. Several trees of undetermined kind grow on the slopes of the otherwise grassy mound. In 1989, when the later picture was made, the mound had become an island in Fort Gibson Lake. Dead standing trees seen in the more recent picture may be those seen growing in the early photograph, replaced now by black willow, a species which is more tolerant of encroachment of the lake. The site is near the west end of a bridge over Fort Gibson Lake on

INDIAN MOUND IN THE GRAND RIVER BOTTOMS

MORE RECENT PHOTOGRAPH

Oklahoma Highway 51. The later photograph was taken near Taylor Ferry Marina.

John Lamon, descendant of the Norman family, and owner of the mound and its surrounding land prior to the lake impoundment, identified the island as being the mound which could be seen in the earlier photograph. A second mound, not visible in either photograph, arises as an elongated island a few hundred yards to the south.

NUMBER THREE

EARLY PHOTOGRAPH

Both a railroad and a motor car bridge spanned the Verdigris River prior to 1935 when the older of these two photographs was made. The picture is part of the Oklahoma Geological Survey Collection at the University of Oklahoma Libraries. The trestles, span, and two piers of the old Missouri, Oklahoma, and Gulf Railroad bridge are in the background. Banks of the river were mostly grass covered at the time of the older picture, with a few trees on the bank to the north of the railroad bridge and a few scattered trees on the slopes to the south.

The "falls" in the river were caused by limestone outcrops which rose above the flow in the stream a short distance below the bridges. Banks of the river had become covered with trees in 1989 when the later picture was made. The railroad bridge had been removed and the

BRIDGE ON THE VERDIGRIS RIVER NEAR OKAY

MORE RECENT PHOTOGRAPH

motor car bridge closed to vehicles. Water from the Kerr-McClellan waterway a short distance downstream from this site has deepened the river and submerged the rocks, obliterating the falls. Tree species on both sides of the river include black willow, cottonwood, American elm, ash, post oak, persimmon, hackberry, and black walnut. The site is a short distance outside the town of Okay in Wagoner County.

NUMBER FOUR

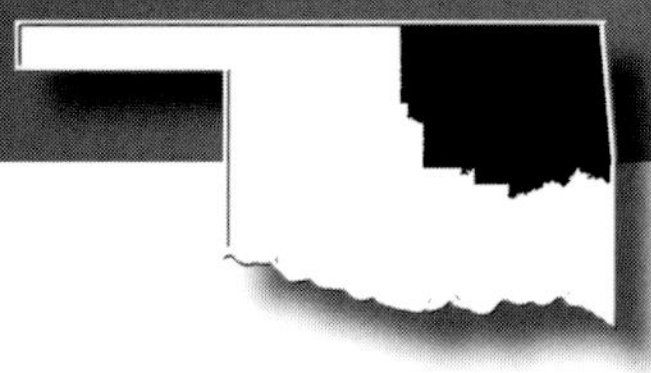

EARLY PHOTOGRAPH

An Oklahoma Geological Survey photograph taken sometime during the 1920s shows a view of a bend in the Arkansas River in Tulsa County as it appears from the top of a hill two miles east of Wekiwa. The winding course with broad, shallow bed, and shifting sandbars is typical of the Arkansas and of other major rivers draining northeast Oklahoma.

Bottomland forests lined both banks of the river except where clearings had been made for pastureland and cultivation, as can be seen in the distant upper left-hand corner of the picture. More than 60 years later, patterns of cleared land on the left bank of the river had changed, as had positions of the sandbars within the stream itself. Otherwise the view of the river appeared much like it did in the 1920s.

ARKANSAS RIVER NEAR WEKIWA

MORE RECENT PHOTOGRAPH

The newer picture was taken from a scenic overlook near exit 29 about 2.5 miles west of Tulsa on the Keystone Expressway. A forest of post oak, black hickory, and elm grew on the river bluffs at the site. The native sandstone enclosure at the overlook is visible in the foreground of the picture. Wekiwa, which consisted of little more than a church and auto repair shop in 1989, is on an unmarked road running parallel to Keystone Expressway south of the river near Northwest 177th Street.

NUMBER FIVE

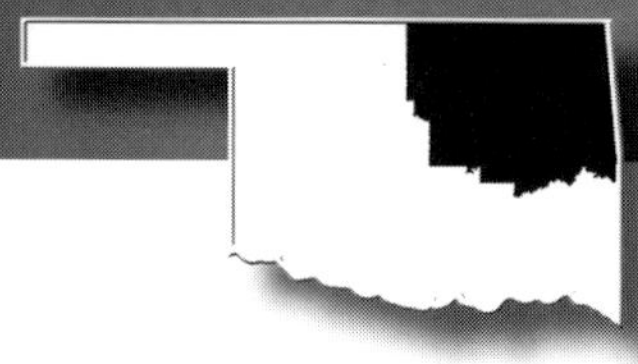

EARLY PHOTOGRAPH

Moccasin Track Bluff is a well-known landmark and picnic site in Creek County near the town of Sapulpa. The bluffs offer an impressive view of the valley beyond Polecat Creek, which runs beneath the bluffs east of the town. An early undated Oklahoma Geological Survey photograph shows an equestrian and her mount near the brink of the sandstone bluffs with an oak-hickory forest growing in the background.

MOCCASIN TRACK BLUFF NEAR SAPULPA

MORE RECENT PHOTOGRAPH

A photograph taken in 1989 includes the same oak-hickory forests in the background. However, growth of trees from below obscured the brink of the bluff along much of its face to the east. Tree species growing along the edge of the bluffs included winged elm, persimmon, honey locust, and blackhaw. The creek bed below the bluffs was littered with trash and debris at the time of the later picture. The site can be reached by traveling along a winding dirt road extending beyond the east end of Thompson Avenue in Sapulpa.

NUMBER SIX

EARLY PHOTOGRAPH

A rider salutes at the approach to an old wooden bridge on the Cimarron River south of the town of Perkins near what is now US-177. The sign warns travelers about a "$5.00 fine for riding or driving over this bridge faster than a walk." Similar penalties were imposed against drivers of more than 15 head of cattle over the wooden span as well as for those defacing the bridge.

The undated photograph from the archives at the University of Oklahoma Libraries was probably made before 1920 when motor vehicles

BRIDGE ON CIMARRON NEAR PERKINS

MORE RECENT PHOTOGRAPH

were scarce. A later photograph from 1987 was made at a point on the river bed where the old bridge had stood. The highway has been relocated and a new bridge replaces the one seen in the older photograph. In the distance, a taller tree stands to the right of the route of the older road whose path has been obscured by typical bottom land woods. Weedy plants, including Johnson grass and sunflowers, can be seen in the foreground of the later photograph.

NUMBER SEVEN

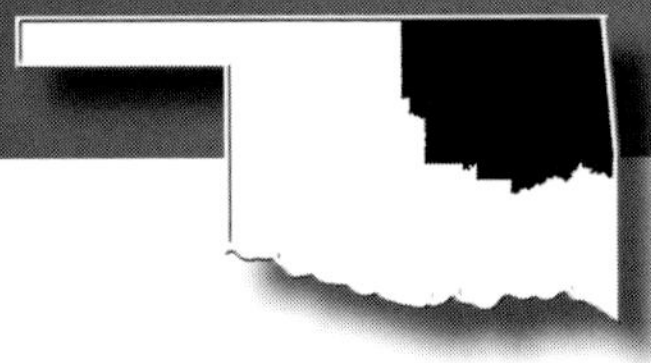

EARLY PHOTOGRAPH

A view of the Medco oil field, two miles east and four miles south of Billings, can be seen in this 1919 photograph by Fred Bullard of the Oklahoma Geological Survey Collection at the University of Oklahoma Library. Oil derricks, storage tanks, and equipment dot the landscape of shale outcrops and grassy flatlands in a scene that is typical of many fields during the early years of the oil industry.

MEDCO OIL FIELD — NOBLE COUNTY

MORE RECENT PHOTOGRAPH

Seventy years later, vegetation on the site remained much as before and consisted of bluestem grasses scattered among the shorter buffalo and grama grasses, with scattered American elm and hackberry trees. Oil production had declined by the late 1980s with only a few producing wells remaining in the Medco field.

NUMBER EIGHT

EARLY PHOTOGRAPH

Courtesy Oklahoma Geological Survey Collection, University of Oklahoma Libraries.

Many of Oklahoma's older wooden and steel road bridges have been replaced by modern reinforced concrete structures. An example is the bridge under construction in 1989 on State Highway 177 over the Chickaskia River in Kay County near Blackwell. Trees and other vegetation which stood on the approaches and in the floodplain in the 1916 photograph were being covered or displaced by the new bridge.

CHICKASKIA RIVER BRIDGE AT BLACKWELL

MORE RECENT PHOTOGRAPH

The river bottom forest of American elm, hackberry, and green ash up and down the stream from the bridge site remained largely intact at the time of the later picture. The new bridge occupies the same site as the previous structure, but its wider approaches and higher roadbed introduce a more imposing element in the landscape than did its smaller predecessor.

SOUTHEAST OKLAHOMA

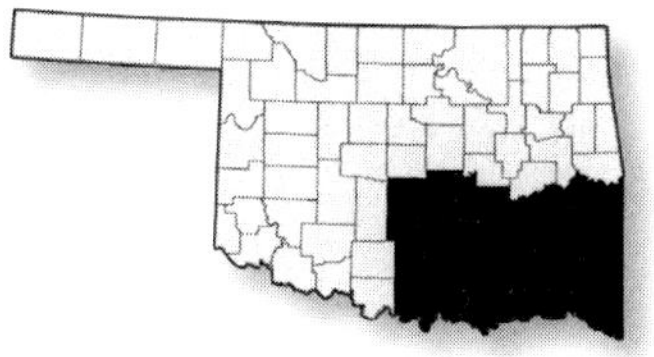

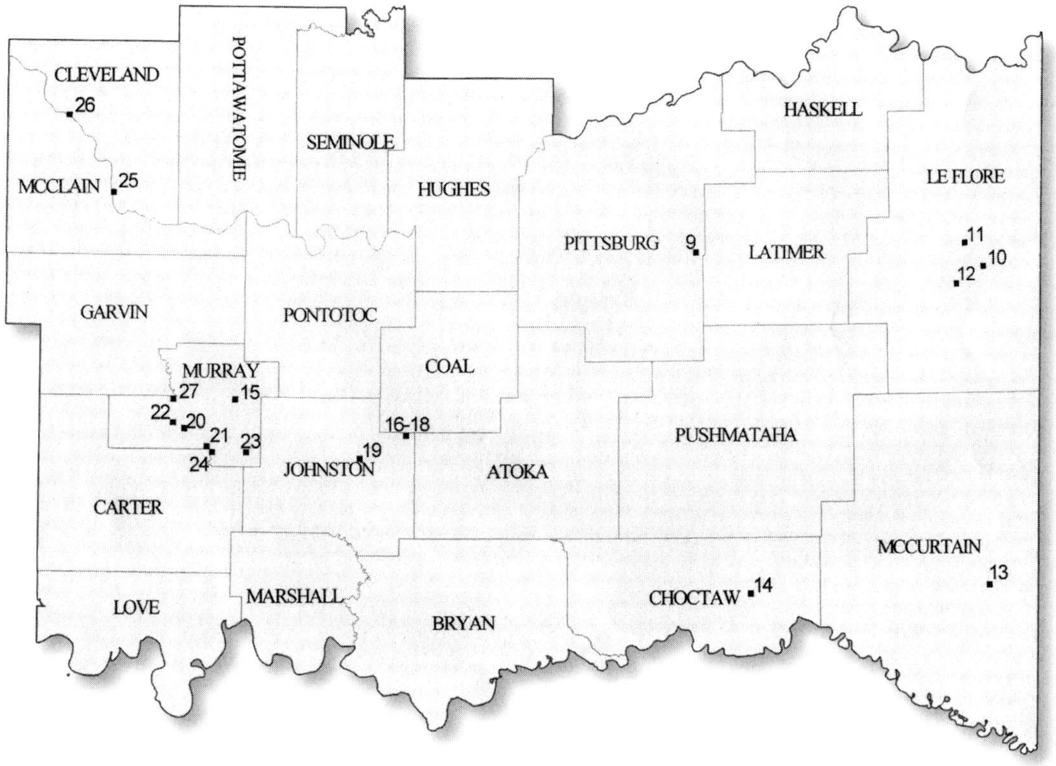

NUMBER NINE

EARLY PHOTOGRAPH

Courtesy Oklahoma Geological Survey Collection, University of Oklahoma Libraries.

Western extremities of the Ouachita Mountains extend into Pittsburg County where remnants of the range form modest ridges near the town of Hartshorne. Belle Starr is one of the westernmost ridges of the mountain chain. This flat-topped ridge lies about one mile east of the town and north of US-270 where it crosses the railroad. The new improved highway runs along the north edge of the mountain.

On the horizon in the older photograph, a pile of coal mine tailings gives the appearance of a mountain. The tailings had been removed or perhaps used in construction of the new highway and are no longer visible in the 1996 photograph.

BELLE STARR MOUNTAIN

MORE RECENT PHOTOGRAPH

The 1945 photograph shows the oak-hickory covered mountain rising above the surrounding grassland, which includes a scattering of trees. A new and larger railroad overpass on US-270 has altered the site. The recent picture taken from the edge of the new overpass looks north toward Belle Starr Mountain. The open prairie of former times had become a woodland consisting mostly of bois d'arc and black locust trees. The oak-hickory forest on the mountain itself appeared to be largely unchanged from previous years.

NUMBER TEN

EARLY PHOTOGRAPH

Courtesy Oklahoma Geological Survey Collection, University of Oklahoma Libraries.

An Oklahoma Geological Survey photograph taken in the early 1900s is of the Kansas City Southern railroad bridge across the Poteau River, south of Heavener, not long after completion of the line. Roadbeds are freshly cleared and vegetation has been removed along the right-of-way. Leafless winter vegetation reveals the rocky overhang on the bank of the river at the right center of the picture. Three unidentified men can be seen near the bridge in the original picture.

RAILROAD BRIDGE NEAR HEAVENER

MORE RECENT PHOTOGRAPH

In 1996, when the newer photograph was made, the bridge and tracks were unchanged from nearly a century before, although vegetation had reclaimed much of what had been denuded right-of-way when the tracks were new. Vegetation also had obscured the rocky river banks which were visible beyond the bridge in the first picture. The rich mixture of trees that had crept back toward the tracks from the surrounding forests included Shumard oak, water oak, mulberry, sweet gum, green ash, American elm, and hackberry, as well as numerous shrubs and vines.

NUMBER ELEVEN

EARLY PHOTOGRAPH

Courtesy Oklahoma Geological Survey Collection, University of Oklahoma Libraries.

The site of the older photograph is about one mile south of Heavener in LeFlore County. The photograph is undated, although the exposed aspects of the road cuts indicate it was taken soon after the railroad was constructed in the early 1900s. Scattered trees above the exposed cuts appear by the pattern of their twisted form to be post oaks.

KANSAS CITY SOUTHERN RAILROAD NEAR HEAVENER

MORE RECENT PHOTOGRAPH

The bare rock seen in the older photograph was heavily overgrown with vegetation in the 1988 photograph. Tree species found around the cut included post oak, blackjack oak, and shortleaf pine. Dense cover on the sides of the cut was composed of greenbrier, honeysuckle, smooth sumac, blackberry, and numerous species of trees, grasses, and forbs.

NUMBER TWELVE

EARLY PHOTOGRAPH

Courtesy Oklahoma Geological Survey Collection, University of Oklahoma Libraries.

The caption on the older of these photographs reads "U.S.G.S. Government bench mark at the township corner, three miles south of Heavener, Ok." The post near the unidentified man marks the corner between Townships 4-5N and Ranges 25-26E in LeFlore County near Heavener. Because of the offset along the First Standard Parallel, there are two locations which fit this description, but the alternative one occurs on a rocky hillside which distinguishes it from this flood plain location.

The area is in the floodplain of the Poteau River and was being cleared for farmland when the pre-1935 photograph was made. A few trees remained from the original floodplain forest. Note the split-rail fence which also includes one strand of barbed wire. The man standing near the center of the picture is not identified. The iron post in the center of the old photograph

TOWNSHIP CORNER SOUTH OF HEAVENER

MORE RECENT PHOTOGRAPH

could not be found at the time of the more recent photograph. The senior author can be seen standing behind a post in the later picture taken in 1988. The post, which had replaced the iron post, stood at a fence corner near where the township corners meet.

The area had not been farmed for many years and a second-growth floodplain forest had reclaimed the site following its abandonment for farming. A few large trees, such as the one to the left of center in the later picture, were found growing in the fence rows, although most of the trees in the area were less than half the diameter of the larger tree. Tree species growing at the site were those typical of the floodplain forest along the Poteau River and included water oak, Shumard oak, sweet gum, black gum, green ash, sycamore, honey locust, hackberry, and basswood.

NUMBER THIRTEEN

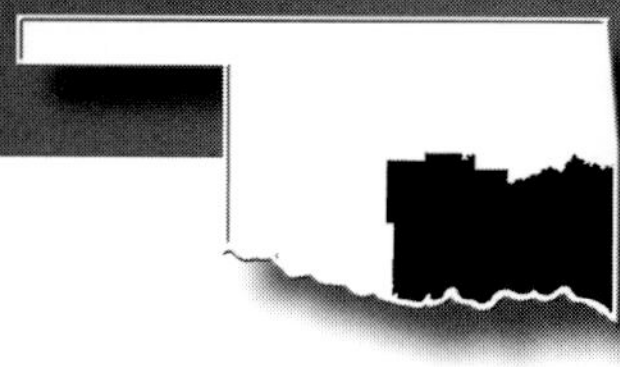

EARLY PHOTOGRAPH

A 1916 photograph by John Cullen of the University of Oklahoma Library is of the giant baldcypress tree on Mountain Fork River in McCurtain County 6.5 miles east of Broken Bow. The photograph captures the massive proportions of this ancient tree. Two unidentified men standing at its base were dwarfed by the eight-foot diameter of its trunk. A bottomland hardwood forest surrounded the tree and vines had climbed several feet up a crevice in the trunk.

GIANT BALD CYPRESS TREE NEAR BROKEN BOW

MORE RECENT PHOTOGRAPH

The 40-foot stump remained in the 1988 photograph of the old baldcypress, although the tree itself was killed by lightning in 1982. Little else appeared to have changed in the floodplain forest surrounding the old tree. The trunk of the old dead tree dwarfs nearby sweet gum, black gum, sycamore, water oak, and American elm. Included among its neighbors were several smaller living baldcypress trees growing in the quiet waters of the swampy creek.

NUMBER FOURTEEN

EARLY PHOTOGRAPH

The view is toward the west from the east side of the river. The automobile in the 1921 photograph belonged to the photographer, Fred Bullard. Professor Bullard, a geologist at the University of Oklahoma, traveled extensively in Oklahoma taking many photographs of the land features of the state. The bridge had a concrete and steel framework with a wooden deck. A solid deck was built of boards laid crosswise, with two lines of boards laid lengthwise to accommodate the wheels of automobiles and wagons.

A few bridges of this type still can be found on back roads in Oklahoma. The road was

KIAMICHI RIVER BRIDGE AT SAWYER

MORE RECENT PHOTOGRAPH

rerouted in 1932 and the old bridge torn down. The photograph taken in 1988 was to the east of the original to better show the forest growth in the background. The fence in this photograph is built in the old roadbed. A few pieces of broken concrete remained from the old bridge. Some effort had been made to keep woody vegetation from growing in the foreground area which was used for pasture. But near the river, the native bottomland forest had become reestablished. Trees visible in the background included green ash, willow, boxelder, water oak, sweet gum, and sycamore.

NUMBER FIFTEEN

EARLY PHOTOGRAPH

Courtesy Oklahoma Geological Survey Collection, University of Oklahoma Libraries.

An early photograph by Murff and Whitehill from the Oklahoma Geological Survey Collection is titled, "Grand Rapids, Platt National Park." The site, also called Little Niagara Falls, is on Travertine Creek near the Visitor Center in the Chickasaw National Recreation Area, formerly known as Platt National Park near Sulphur. The photograph shows the creek in near natural condition, although some trees had been cut down in the center background.

LITTLE NIAGRA FALLS

MORE RECENT PHOTOGRAPH

A photograph taken in 1988 shows the changes brought about by conversion of the site to a popular recreation area. The opposite side of the stream from which the photographs were taken had been filled, leveled, and landscaped for a picnic area. Various concrete structures had been placed in the stream and along its banks to deepen pools for swimming and to stabilize the banks. An area on the near side of the creek remained in native bottomland forests including bur oak, Shumard oak, American elm, hackberry, and several other kinds of trees.

NUMBER SIXTEEN

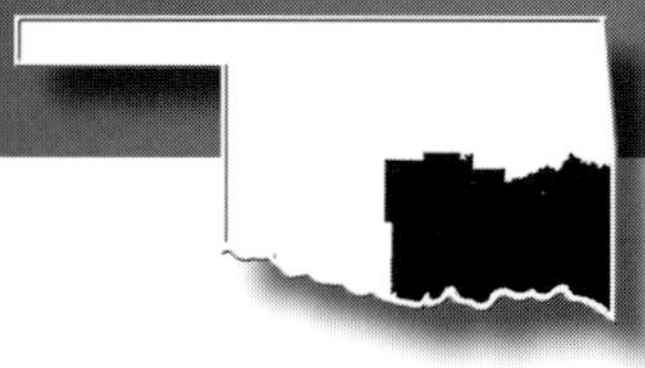

EARLY PHOTOGRAPH

Courtesy Oklahoma Geological Survey Collection, University of Oklahoma Libraries.

A young cyclist views the mineral springs at Bromide, a town on the line between Coal and Johnston counties. These springs were popular in the early years of the twentieth century because of their presumed medicinal properties. The disagreeable taste of the spring water was thought to be an indication of its beneficial qualities.

Local residents improved the spring area with rock walls, pools, roads, and benches in the hope of developing Bromide into a health resort. A 1916 photograph by John Cullen shows one of the popular springs which emerged about halfway between the buttress of the retaining wall. When the use of mineral water was no longer popular, visits to the springs

SALT BROMIDE SPRINGS

MORE RECENT PHOTOGRAPH

at the town were largely discontinued. The springs were abandoned and many of the stone works surrounding them became enshrouded in vegetation.

Everything connected with the springs, except for some of the stonework, has disappeared. The springs still flowed with the same unpleasant tasting water as before. The bois d'arc tree visible just behind the buttress at the right appeared to be the same tree that could be seen in the earlier picture. Vegetation apparently had changed very little in the seven decades between the two pictures and consisted of an open woodland of post oak, chinquapin oak, bois d'arc, and hackberry with a ground cover of grasses.

NUMBER SEVENTEEN

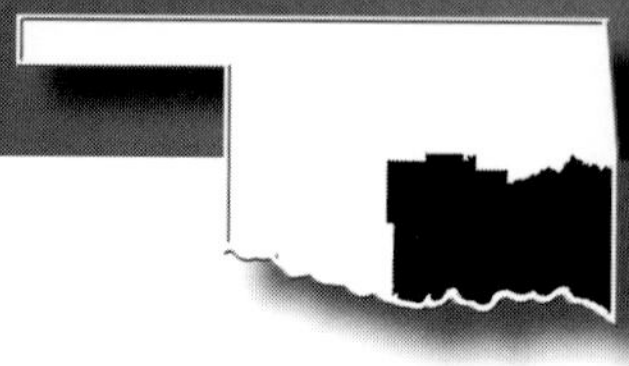

EARLY PHOTOGRAPH

Courtesy Oklahoma Geological Survey Collection, University of Oklahoma Libraries.

A 1916 photograph shows one of the more elaborate springhouses at Bromide, consisting of a system of retaining walls built from rough stone. One spring issues from a basin in the lower center of the picture and one or more springs were inside the springhouse. By 1988, all of the retaining walls had crumbled and the excavation had silted in. However, the solidly-built springhouse was intact.

SWEET BROMIDE SPRINGHOUSE

MORE RECENT PHOTOGRAPH

The springs apparently still flowed underground as evidenced by the wet soil and dense vegetation in the foreground and near the springhouse. Tree and other plant species including black willow and cattails were similar to those found in other wet sites in the vicinity. The older photograph is by John Cullen.

NUMBER EIGHTEEN

EARLY PHOTOGRAPH

This John Cullen 1916 photograph shows a pipe emerging from the spring, apparently for the purpose of allowing water from the spring to be pumped for use at another location.

Courtesy Oklahoma Geological Survey Collection, University of Oklahoma Libraries.

SPRING AT BROMIDE

MORE RECENT PHOTOGRAPH

A photograph taken in 1988 shows that the solidly-constructed stonework at the spring remained mostly intact. At the time of the later picture, the site was difficult to find because it was almost completely hidden by vegetation.

NUMBER NINETEEN

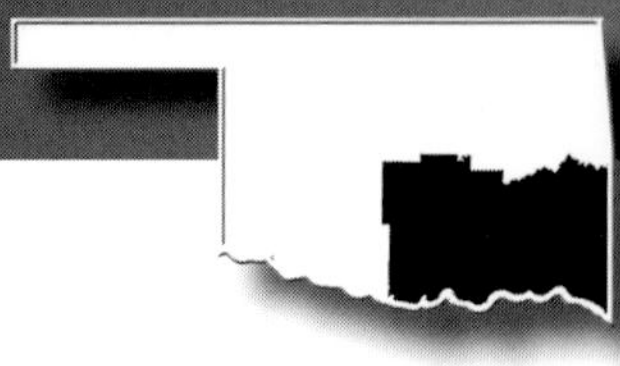

EARLY PHOTOGRAPH

Courtesy Oklahoma Geological Survey Collection, University of Oklahoma Libraries.

Big Blue River in Johnston County divides and forms several low falls as it flows over a granite ledge west of Wapanucka. The one shown here is probably the westernmost, as it was the only one having a vertical drop at the time of the more recent photograph in 1988. The older photograph, taken before 1935, may have been made from a boat, because at the time of the more recent photograph, a large, deep pool extended completely across the river below the ledge and includes the camera position of the earlier photograph. Vegetation appears to be the typical bottomland forest type.

BIG BLUE RIVER FALLS

MORE RECENT PHOTOGRAPH

The later photograph was taken from the riverbank and shows the width of the pool as it appeared in 1988. Appearance of the falls had changed from earlier times. The later photograph was taken during a time of drought, as is indicated by the low flow of water in the falls. Vegetation appeared to be similar in the two pictures. Trees visible were typical of the bottomland forest in the area including bur oak, Shumard oak, American elm, and green ash. Growing nearby in the rocky banks of the Blue River, though not visible in either photograph, are seaside alders, unusual trees that are found only on the Blue River in Oklahoma and along streams near the coast of Delaware and Maryland.

NUMBER TWENTY

EARLY PHOTOGRAPH

Courtesy Oklahoma Geological Survey Collection, University of Oklahoma Libraries.

An early undated photograph shows the wooded setting of Price's Falls in the Arbuckle Mountains.

PRICE'S FALLS

MORE RECENT PHOTOGRAPH

The more recent 1988 photograph shows a wooded background that was largely unchanged, although the surrounding forests appear to include more and larger trees than in the earlier scene. Tree species growing in the area are typical of the bottomland habitat. Horsetail is conspicuous in the understory. Water milfail is abundant in the waters of the creek. Price's Falls is located about five miles southeast of the junction of Oklahoma OK-770 and US-77.

NUMBER TWENTY-ONE

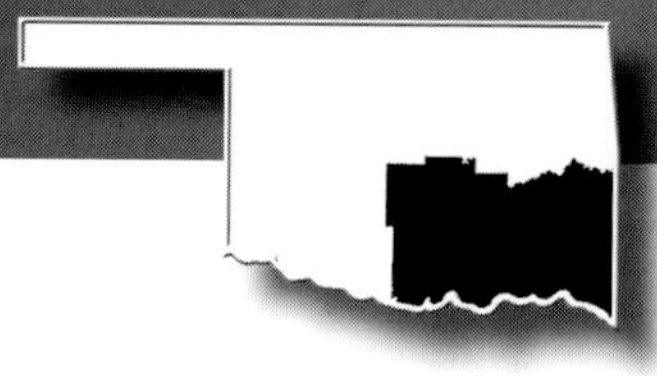

EARLY PHOTOGRAPH

Courtesy Oklahoma Geological Survey Collection, University of Oklahoma Libraries.

The site at which these photographs were made is on Atcheson, Topeka, and Santa Fe Railroad about four miles south of the town of Dougherty in Murray County. Upland vegetation on these rocky limestone hills, which are part of the Arbuckle Mountains, is open woodland of small rock cedar trees and shrubby oaks. Some of the shrubs in the right foreground are recognizable as short-lobed oak which are locally abundant in the Arbuckles.

The floodplain forest across the river to the right is sparsely vegetated, probably because of the shallow rocky soil. The place from which this pre-1935 photograph was taken has since been overgrown by dense shrubs and trees. In 1988, it was no longer possible to make a photograph from the exact location at which the earlier picture was taken. The thick and

WASHITA RIVER GORGE

MORE RECENT PHOTOGRAPH

brushy vegetation on the slopes above the river consisted mostly of short-lobed oak, aromatic sumac, and rock cedar, the same species from earlier times. Little bluestem, yucca, and various forbs occurred on the few remaining openings in the area.

Upland vegetation had changed in the nearly half century between photographs from a savanna-like appearance to that of a shrub land, as a result of an increase in the number of short-lobed oaks and aromatic sumac. Bottomland forests near the river itself were composed largely of cottonwood, green ash, and black willow and appeared to be essentially unchanged. However, behind the camera station lay a large, abandoned stone quarry which severely impacted the contour and vegetation of the area.

NUMBER TWENTY-TWO

EARLY PHOTOGRAPH

Courtesy Oklahoma Geological Survey Collection, University of Oklahoma Libraries.

A 1928 photograph captured Turner Falls at a season of high water and reveals the falls and Honey Creek below the falls in their natural state. The bottomland forests that dominated the canyon appeared to be largely undisturbed as did the juniper-oak woodland dominated by rock cedar and short grass openings on the surrounding hills. The major change visible in the more recent photograph is the construction of a low dam in the basin beneath the falls and the clearing

TURNER FALLS

MORE RECENT PHOTOGRAPH

of most of the trees along the banks of the creek where it flows out of the pool. A road had been constructed along the edge of the stream. Common trees in the canyon are American elm, green ash, and chinquapin oak. Forest and grassland on the slopes above and surrounding the falls and creek remained largely unchanged from 60 years before, although trees had increased both in size and number during the interval between the two pictures. Both the 1928 and 1988 photographs were made looking south toward the falls from near the observation point and scenic turnout of US-77 about 6.5 miles south of the town of Davis.

NUMBER TWENTY-THREE

EARLY PHOTOGRAPH

Courtesy Oklahoma Geological Survey Collection, University of Oklahoma Libraries.

Several limestone formations are tilted almost vertically in the Arbuckle Mountains in northern Carter County and have weathered into rows of standing rocks, giving the orderly appearance of tombstones. The site of these photographs is one-half mile south of the Carter-Murray County line on US-77. The earlier photograph was made in 1939 and the later in 1989.

TOMBSTONE FORMATIONS

MORE RECENT PHOTOGRAPH

Both pictures were taken in the summer and show scattered trees and shrubs in the grassland. Grasses were unusually tall in the summer of 1989 because of abundant rainfall, obscuring the smaller rocks of the formation. Except for these taller grasses, vegetation appears to be essentially unchanged after 50 years. Trees present at the site were netleaf hackberry and American elm. Shrubby vegetation was mostly winged sumac with dominant grasses being little blue stem and sideoats grama.

NUMBER TWENTY-FOUR

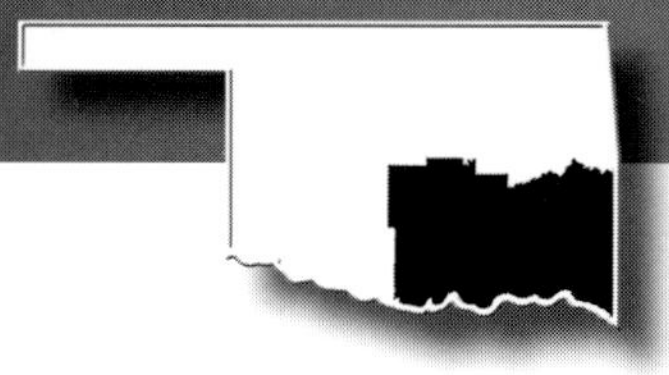

EARLY PHOTOGRAPH

Courtesy Oklahoma Geological Survey Collection, University of Oklahoma Libraries.

An early undated photograph of White Dome, or White Mound as it is named on some maps, shows fossil hunters at this well-known excavation site near Dougherty in Murray County. The mound itself is almost devoid of vegetation, but is surrounded by wooded and grass-covered rolling hills.

WHITE DOME

MORE RECENT PHOTOGRAPH

A 1988 photograph shows White Dome largely unchanged from its appearance in the earlier photograph. However, the grassy hill in the right background of the scene had been capped with trees. Some of the trees had spread into low places near the mound itself. Tree species present in the vicinity included rock cedar, chinquapin oak, and soapberry, as well as several kinds of shrubs. Little bluestem dominated the grasslands, mixed with verbena, big flower coreopsis, and other common forbs.

NUMBER TWENTY-FIVE

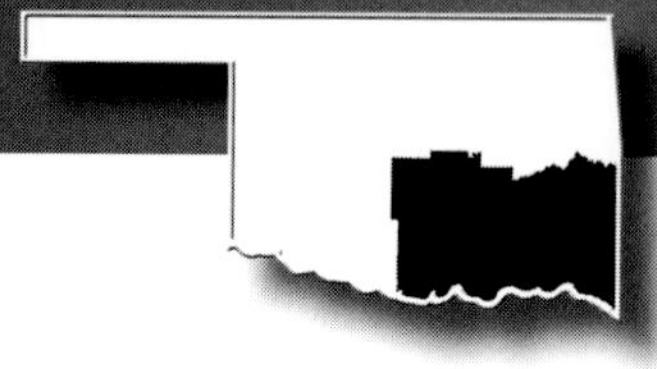

EARLY PHOTOGRAPH

Courtesy E.A. Coleman Collection, University of Oklahoma Libraries.

An old photograph of a bridge across the Canadian River between Lexington in Cleveland County and Purcell in McClain County was made about 1900. A photograph made in 1988 reveals many changes, both in the natural surroundings of the river and in bridges that have spanned the stream.

The old suspension bridge seen in the early photograph joins the bank of the river near its first pier, and rises only a few feet above the course of the stream. Floods had removed the entire structure, and most of its successors, with only one steel pier of a more recent span remaining. The new bridge seen to the left in the later photograph was constructed in 1938

LEXINGTON BRIDGE

MORE RECENT PHOTOGRAPH

at a sufficient height to escape the force of the floods. In 1968 it was named the James C. Nance Bridge in honor of the legendary legislator and newspaper publisher from Purcell. The stream itself had widened at this point in its course, but the larger amount of water visible in the more recent photograph is probably a result of seasonal differences at which the earlier and later pictures were taken. Both photographs show a tree-lined west bank of the river in the distance, although the open canopy provided by large American elm and pecan trees has been replaced in the more recent photograph by willow and cottonwood. On the near side of the river, bare sand had become covered with grass and was being invaded by a few woody plants.

NUMBER TWENTY-SIX

EARLY PHOTOGRAPH

Courtesy Oklahoma Geological Survey Collection, University of Oklahoma Libraries.

The graceful arches of a steel bridge fade into the distance in this early undated photograph. The photograph was taken near the road that is now OK-9 at the Canadian River south of Norman in Cleveland County. In 1988, a single concrete pier at the south edge of the river is all that remained of the old steel bridge. That pier was joined by two pairs of concrete supports. The bridge had become a victim to the currents of the river.

CANADIAN RIVER BRIDGE AT NORMAN

MORE RECENT PHOTOGRAPH

Both bridges had been replaced by a more substantial concrete bridge on I-35 a few hundred yards to the west. The river channel seemed much narrower than in earlier times. In the more recent picture, a nearly solid phalanx of trees is clearly visible on the south bank of the river as seen from the same vantage point at which the earlier photograph was taken.

Cottonwoods and willows were common trees found near the river. An unidentified boy stands by a background of sandbar willow on the north bank in the older photograph. Except for seasonal changes, the north bank was similar at the time of both photographs, with scattered willows being the only plants established in the shifting sands of the riverbed.

SOUTHWEST OKLAHOMA

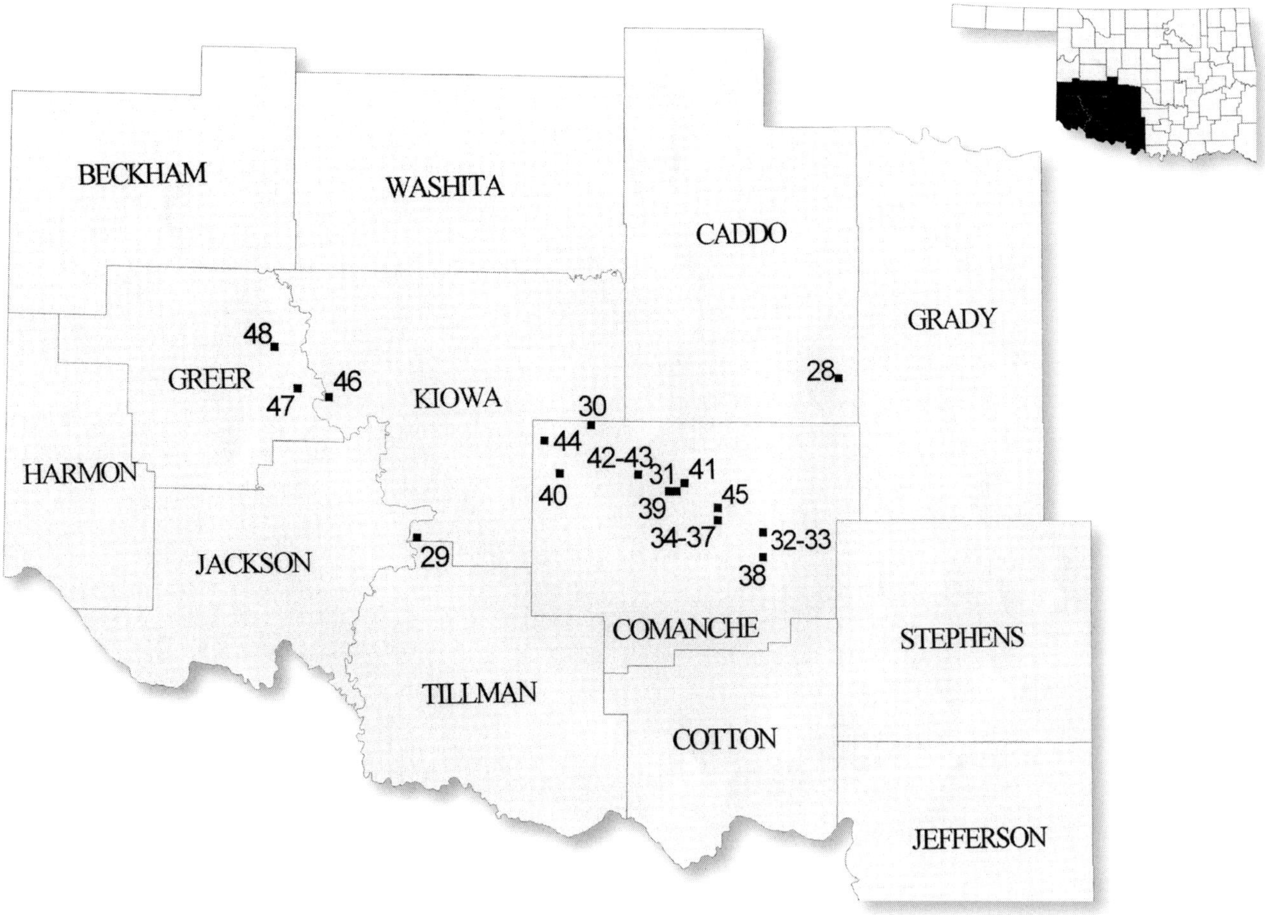

NUMBER TWENTY-SEVEN

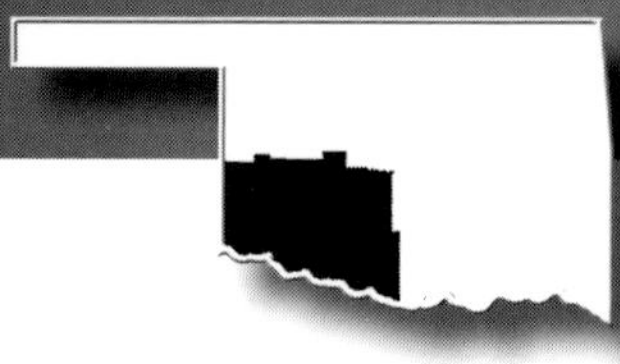

EARLY PHOTOGRAPH

Courtesy Oklahoma Geological Survey Collection, University of Oklahoma Libraries.

A rough limestone column on the line between Murray and Garvin counties marks the Initial Point of the General Land Office survey of the area. The column remains in place and was unchanged in 1988 at the time of the newer of these two photographs. The pioneer geologist, Charles N. Gould, stands beside the monument in the older photograph dated "before 1935."

This monument, erected about 1870, bears the letters "I.P." on the west side, and "I. Mer." on the east side. The marker identifies the southern most point of the Indian Meridian, which along with Base Line, are the east-west and north-south coordinates for the land survey of the state of Oklahoma except for the Panhandle.

Rock cedars and short grasses continued to surround the marker at the time of the newer photograph. However, heavy grazing and the suppression of fire had resulted in an increase in the number and size of cedars on the site.

INITIAL POINT OF THE OKLAHOMA LAND SURVEY

MORE RECENT PHOTOGRAPH

A narrow trail through the trees had been bulldozed east along the fence and a wider swath cleared to the north of the monument.

The site is one mile south of OK-7, seven miles west of the town of Davis. A historical marker on the highway at that point calls attention to the location of Fort Arbuckle, the administrative center for the area when the land survey began in 1870. A steep-sided canyon with an intermittent stream and a permanent pool lies in the path halfway between the highway and the Initial Point monument. The senior author stands in the same position beside the marker that was occupied by Gould in the earlier photograph.

NUMBER TWENTY-EIGHT

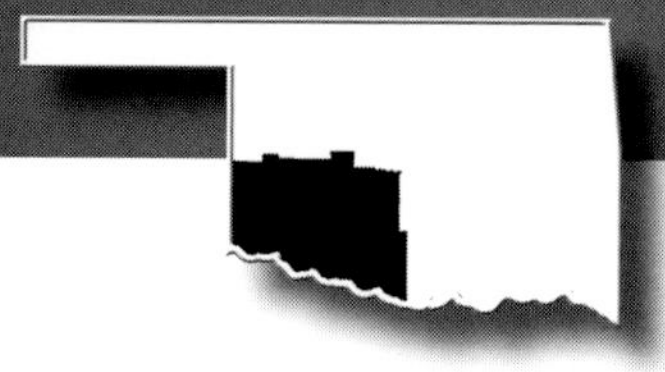

EARLY PHOTOGRAPH

This hill with a flagpole is located at the corner of Fifth Avenue and D Street in the town of Cement in southeastern Caddo County. No woody vegetation is visible in the 1941 photograph, and much of the gravelly soil below the sandstone cap is exposed. In this Oklahoma Geological Survey Collection photograph, two

HILL WITH FLAGPOLE

MORE RECENT PHOTOGRAPH

climbers are at the right of the summit. Several small American elms and black locust trees are visible in a later picture taken in 1988. There appeared to have been an increase in herbaceous plant cover in the 47 years between photographs, especially to the left of the footpath.

Herbaceous plants present were those typical of mixed grass prairie, particularly little bluestem and grama grasses.

NUMBER TWENTY-NINE

EARLY PHOTOGRAPH

Courtesy Oklahoma Geological Survey Colllection, University of Oklahoma Libraries.

The rocky bed of Beaver Creek was almost bare of vegetation in the 1951 photograph. By 1990, the old wooden span had been replaced by a more substantial concrete bridge with paved approaches several feet higher than its predecessor. The streambed had been

BEAVER CREEK BRIDGE

MORE RECENT PHOTOGRAPH

invaded by cottonwood trees, black willow, and American elm, as well as switch grass and a variety of weedy plants growing in soil eroded into the creek bed from surrounding fields. The site is on a section line road about three miles west of Marlow in Stephens County.

NUMBER THIRTY

EARLY PHOTOGRAPH

Courtesy Oklahoma Geological Survey Collection, University of Oklahoma Libraries.

Saddle Mountain presents a distinctive profile above the flat grasslands in this photograph taken about 1935. The mountain in Comanche County is located two miles east and two miles south of the near-abandoned town of Saddle Mountain, a short distance across the line in Kiowa County in southwestern Oklahoma.

SADDLE MOUNTAIN

MORE RECENT PHOTOGRAPH

Blackjack-post oak forests on the mountain remained unchanged in 1990. Prairie vegetation continued to dominate the flatlands surrounding the mountain.

NUMBER THIRTY-ONE

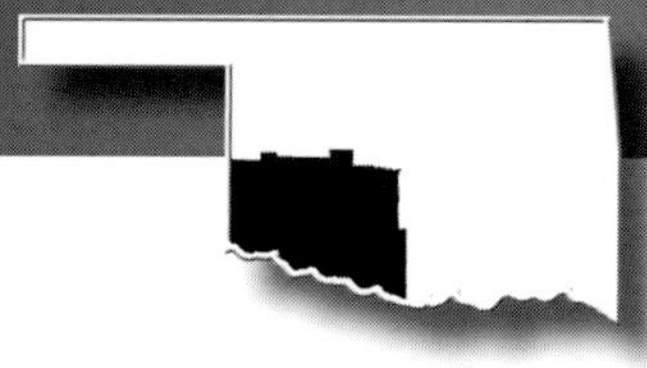

EARLY PHOTOGRAPH

Courtesy Oklahoma Geological Survey Collection, University of Oklahoma Libraries.

An undated early photograph shows a pair of heavily-clothed "mermaids on Medicine Park Rapids," as the caption reads on this photograph. Aside from the mermaids, there is no evidence of human activity at the creek. The vegetation on the rocky hillside at the left appears to be much the same as that seen in the Wichita Mountains at the time of the more recent photograph.

The 1988 photograph shows the changes in the creek. The high dam in the background was built in 1939 and confines Lake Lawtonka,

MEDICINE CREEK

MORE RECENT PHOTOGRAPH

an important source of water for the towns of Lawton and Fort Sill. The rapids are covered by the pool behind one of several low dams which have been built on the creek in Medicine Park.

Vegetation on the hillside appeared to be unchanged and consisted of small post-oak and blackjack oak trees with little bluestem, grama grasses, and other grasses and forbs typical of the mixed-grass prairie.

NUMBER THIRTY-TWO

EARLY PHOTOGRAPH

The description of this photograph by William Soule reads, "East line of officers quarters at Ft. Sill under construction in 1870." Two houses in the middle of the scene appear to have been completed, and others were in various stages of construction at the time. No trees are visible in the early picture because the quarters were being constructed on a grassy hill where trees did not usually occur at the time the post was built. The photograph is from the Ft. Sill Museum collection.

OFFICERS QUARTERS AT FORT SILL

MORE RECENT PHOTOGRAPH

The same buildings were being used as officers quarters in the late 1980s when the more recent picture was taken, although they had been enlarged and remodeled several times. The streets were lined with large oaks and elms and the old parade ground at the right of both photographs was still used for ceremonies.

NUMBER THIRTY-THREE

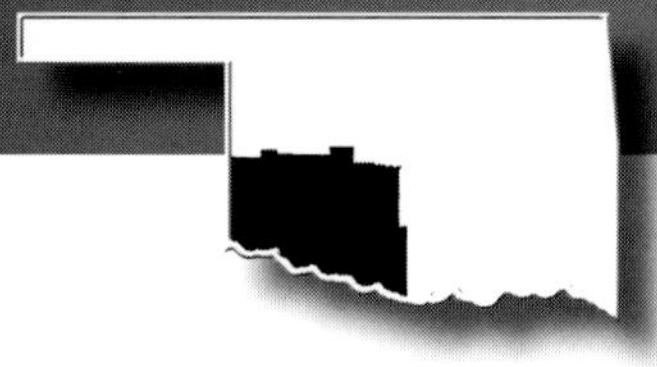

EARLY PHOTOGRAPH

The earlier of these two photographs is an undated selection from the Fort Sill Museum of a battery review ceremony on Pritchard Field sometime before 1920. Howitzers were pulled by horses. The south side of the Medicine Bluffs is visible in the background behind a row of recently-constructed officers' quarters on what was called the "New Post." Native prairie vegetation appears to cover the parade ground.

PRITCHARD FIELD AT FORT SILL

MORE RECENT PHOTOGRAPH

The later photograph was made in 1989 and shows many changes over a period of 70 years or more. Trees had been planted on the streets around the field and had grown to large size. The houses from the earlier picture remain standing with more homes constructed. Native grasses on the field, which was used as an athletic field, had been replaced by Bermuda grass.

NUMBER THIRTY-FOUR

EARLY PHOTOGRAPH

These photographs were taken on the level ground and gentle slopes south of Medicine Bluffs at Fort Sill. The earlier photograph from the Fort Sill Museum Collection shows a heavy howitzer and its crew in 1919. At that time the area was grassland with only a few small shrubs on the hillside. Visitors to the area in the mid-1800s remarked that the south side of the bluffs was covered with grass.

MOTORIZED ARTILLERY AT FORT SILL

MORE RECENT PHOTOGRAPH

The 1989 photograph shows a large increase in woody vegetation in this area during the previous 70 years. Nearly all of the woody vegetation on the level ground between the road and the hill was mesquite, while most of the trees in the saddle between the two hills were American elm and netleaf hackberry.

NUMBER THIRTY-FIVE

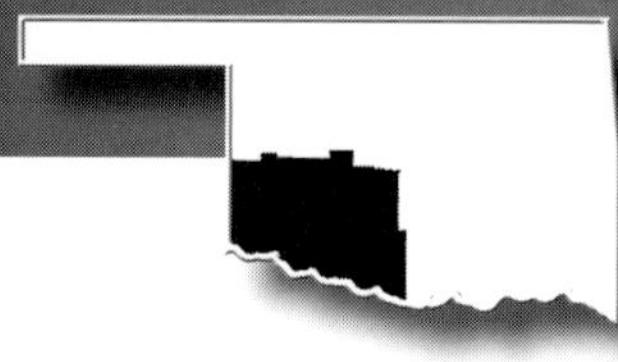

EARLY PHOTOGRAPH

Courtesy Fort Sill Museum.

The earlier photograph was taken around 1875 and is labeled, "Cavalry at Medicine Bluff and Creek, Ft. Sill, I. T." The creek bottom was heavily wooded, except on the south bank where floodwaters prevented the growth of most vegetation. The bluffs were mostly bare except for a few small trees and shrubs.

CAVALRY AT MEDICINE BLUFF AND CREEK

MORE RECENT PHOTOGRAPH

The scene in the 1989 photograph appears much the same. The creek is narrower, possibly because upstream dams had reduced the volume and velocity of floodwater. Common trees in the creek bottom were bur oak, southern hackberry, American elm, and soapberry. The smaller trees on the bluffs were eastern red cedar and American elm.

NUMBER THIRTY-SIX

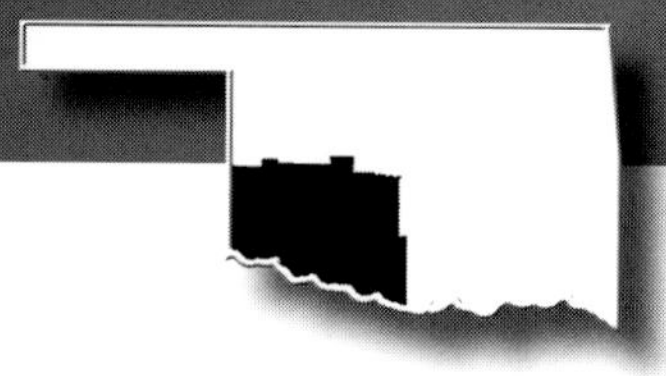

EARLY PHOTOGRAPH

Courtesy Fort Sill Museum.

The earlier photograph was taken by William Soule in 1871 and shows Medicine Bluff Creek and the east end of the Medicine Bluffs. The picture probably was made in late summer or autumn because some of the trees seem to have partially lost their leaves. The creek bed itself appears dry.

MEDICINE BLUFF CREEK

MORE RECENT PHOTOGRAPH

The later photograph was made in 1989. Vegetation on the bluffs and on the floodplain appeared to be essentially unchanged in the 118 years between photographs. The only significant difference was the presence of standing water seen in the later photograph, the result of a small dam downstream.

NUMBER THIRTY-SEVEN

EARLY PHOTOGRAPH

This 1870 William Soule photograph, perhaps the oldest photograph in the Fort Sill Museum Collection, is of Cavalry Gap in the Medicine Bluffs at Fort Sill. The photograph includes the darkroom tent of the pioneer photographer on the bank of the creek. Three soldiers are perched on the large rock in the center of the picture and others are scattered among the trees and rocks nearby.

Courtesy Fort Sill Museum.

CAVALRY GAP

MORE RECENT PHOTOGRAPH

In the older photograph, both black and white troops are present, evidence of what may have been early racial integration in the United States Army. In spite of heavy use of the gap as a training area, vegetation appears to be much the same in the 1989 photograph.

A trail passes through the gap, and many of the larger rocks had been removed or covered. Coniferous trees visible in both pictures are eastern redcedar. Deciduous trees are American elm.

NUMBER THIRTY-EIGHT

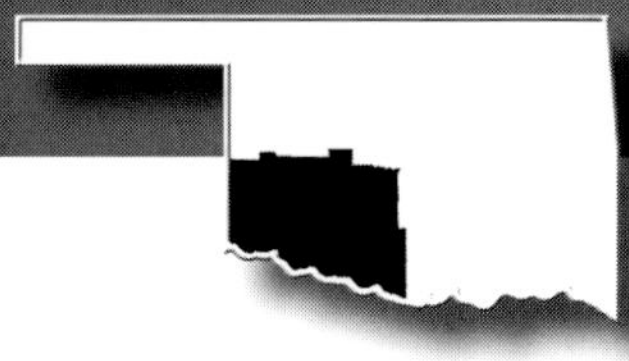

EARLY PHOTOGRAPH

A photograph from the Fort Sill Museum Collection, taken sometime between 1901 and 1907, shows one of several prisoner-of-war camps built to house Geronimo's band of Apaches who were held at Fort Sill from 1894 to 1913. The area is north of the Medicine Bluffs, visible in the distance.

APACHE P.O.W. VILLAGE

MORE RECENT PHOTOGRAPH

The 1989 photograph was taken from a slightly different location because the original site is now covered by woody vegetation. In the more than 80 years between the two photographs, trees had spread from the ravine behind the buildings up the low hills in the background and toward the camera location. The shrubs in the foreground of the more recent photograph are Chickasaw plum and smooth sumac. The trees in the 1989 photograph are mostly post-oak and American elm. Most of those in the older photograph appear to be post oaks.

NUMBER THIRTY-NINE

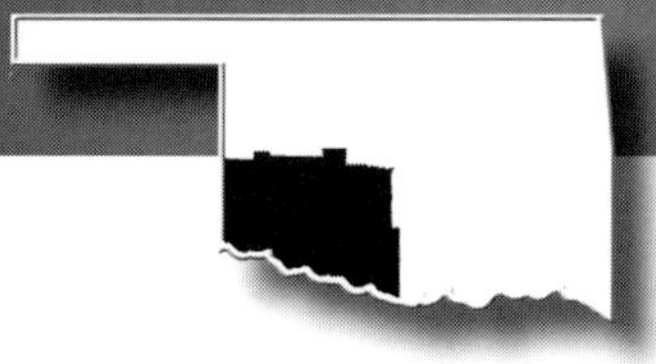

EARLY PHOTOGRAPH

Courtesy Wichita Mountains Wildlife Refuge Collection.

An archway built of native stone around 1920 announces the approach to the Wichita Mountains Wildlife Refuge. This entrance is at the east edge of the refuge on OK-49 and provided access to the refuge until 1969 when the highway was straightened and a new entrance was created a few hundred yards to the north. A fence closed the entrance, and the old roadway was abandoned.

OLD STONE GATE

MORE RECENT PHOTOGRAPH

The native blackjack-post oak forest had reclaimed much of the old roadway as can be seen in the 1997 photograph. Native grasses and forbs covered much of the roadway leading to the old gate. Beyond the gate, trees had become reestablished. Dense woods also had established in the area south of the new gate, making the old gate difficult to see from the new highway.

NUMBER FORTY

EARLY PHOTOGRAPH

Courtesy Wilson Collection, University of Oklahoma Libraries.

This 1906 photograph is of the Bonanza Mine and Smelter in the Wichita Mountains Wildlife Refuge. The mine was on the lower slopes of Mount Lincoln and can be reached by a trail along Fawn Creek from Indiahoma Road.

BONANZA MINE

MORE RECENT PHOTOGRAPH

All evidence of the mine, smelter, and conveyor had disappeared in 1990, except for the mine shaft itself and a small pile of tailings. The original cross timbers type of vegetation had reclaimed the site. Bluestem grasses and prickly pear cactus occurred in the openings within the blackjack-post oak forest. The site is within the Charon's Garden Wilderness Area of the refuge.

NUMBER FORTY-ONE

EARLY PHOTOGRAPH

Courtesy Western History Collection, University of Oklahoma Libraries.

This 1870 view in the Wichita Mountains is labeled, "Pacer's Camp near Mt. Scott." The photograph by W. S. Campbell was taken near Medicine Bluff Creek at a location that later was to be covered by the waters of Lake Lawtonka.

MOUNT SCOTT

MORE RECENT PHOTOGRAPH

The 1989 view is from the eastern shore of the lake. Both photographs were made in winter. In the older picture, the area below the first large rock outcrop appears to be densely covered with trees. The same covering is present 119 years later. Trees on Mount Scott were mostly post-oak and blackjack oak on the lower slopes, with eastern red cedars increasing in number at higher elevations.

NUMBER FORTY-TWO

EARLY PHOTOGRAPH

Courtesy Fort Sill Museum.

"Lt. Samuel Woodward, adjutant of the 10th Cavalry, hunting elk near Mt. Sheridan, 1870," is the caption on this photograph. The area appears to have been burned so that boulders and rocky outcrops are more prominent than in the 1989 photograph. While the exact location of Lt. Woodward's position could not be found at the time of the later picture, the vicinity is easily recognizable from the distinctive profile of Mount Sheridan and its conical companion Mount Roosevelt on the left.

Numerous houses and several roads had

MOUNT SHERIDAN

MORE RECENT PHOTOGRAPH

changed the appearance of the foreground. In the older picture, oak savanna can be seen to extend over the level and rolling terrain at the base of the mountain, while the lower slopes of the mountain were almost bare. A few small trees are visible on the upper slopes. At the time of the later photograph, the level and rolling area at the base of the mountain was still covered with oak savanna, but the mountain slopes, except for outcrops of solid granite, were covered thickly with small post-oak, blackjack oak, and eastern redcedar.

NUMBER FORTY-THREE

EARLY PHOTOGRAPH

Mount Sheridan and its cone-shaped neighbor, Mount Roosevelt, present a distinctive profile among the numerous peaks within the Wichita Mountains Wildlife Refuge. The wildlife refuge was established around 1900 as part of the national forest system. Trees were planted within the refuge in keeping with the forest service's mission.

A 1912 Frank Rush photograph shows a crew planting eastern redcedar trees in the grassland east of the mountains a few years after the refuge was established. Two photographs in 1997 near the same site reveal dramatic changes that had taken place since the time of the plantings. One of these two more recent photographs was taken in the grasslands with the eastern redcedar

TREE PLANTING AT MOUNT SHERIDAN SW 43

MORE RECENT PHOTOGRAPH

Courtesy Wichita Mountains Wildlife Museum.

plantation to the back of the camera. In this photograph, blackjack-post oak forests can still be seen growing at the foot of the mountains, while native eastern redcedar trees had become established nearly as far as the mountain peaks. The second of the two more recent pictures taken further east shows rows of eastern red cedar trees which by 1997 completely obscured the two peaks. Mount Sheridan and Mount Roosevelt are near the north edge of the refuge and can be seen from the road leading to the town of Meers.

NUMBER FORTY-FOUR

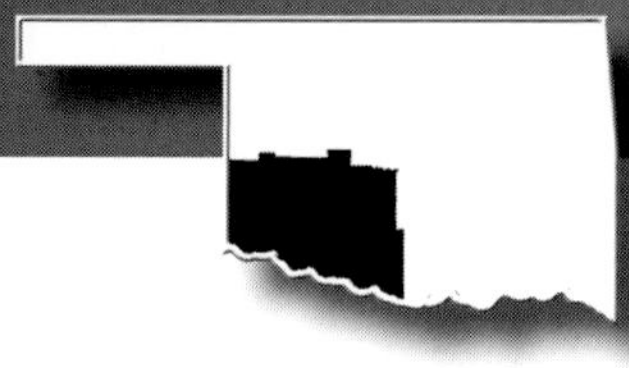

EARLY PHOTOGRAPH

Wichita Mountains Wildlife Refuge in Comanche County was originally established as a part of the national forest system shortly after 1900. This 1911 United States Forest Service photograph, with Baker Peak in the background, shows a team of workers planting honey locust seedlings in the grasslands near the peak in an attempt to fulfill the policy of encouraging the growth of trees within the forest service system. Cattle seen grazing near the base of Baker Peak may be free-ranging stock from ranches nearby.

In 1990, when the recent photograph was

BAKER PEAK

MORE RECENT PHOTOGRAPH

taken, numerous small locust trees were present in the area along with one older large tree about four inches in diameter which may have descended from the original planting. Other larger trees had become established on the slopes approaching the mountain, including American elm, bois d'arc, post oak, blackjack oak, and eastern redcedar. The grasslands in the foreground include the common bluestem grasses, Indian grass, and a number of native forbs.

NUMBER FORTY-FIVE

EARLY PHOTOGRAPH

Courtesy Fort Sill Museum.

The site near Punch Bowl Road on the Fort Sill military reservation is one-half mile north of the Medicine Bluffs facing northwest toward Mount Scott, visible in the center background. The original photograph, taken between 1894 and 1913, is from the Swett Collection in the Fort Sill Museum. Several buildings of one of the Apache prisoner-of-war villages are barely visible in the distance and to the left of a cultivated field. The cultivated field is recognizable by the straight lines of its edges.

The older picture and one made in 1989 show large areas of both woodland and grassland, but it is obvious that the areas of woodland had

EAST END OF THE WICHITA MOUNTAINS

MORE RECENT PHOTOGRAPH

increased considerably in the more than 80 years between photographs. Trees in the older picture were growing mostly along the ravines, with isolated trees and small groves on the upland between the ravines. By 1989, the trees had spread outward from the drainage system and covered much of the level upland. Dominant tree species in the area in 1989 were post oak and blackjack oak, typical of the Cross Timbers and Oak Savanna vegetation types. Those visible in the older photograph also appear to be post oak and blackjack oak.

NUMBER FORTY-SIX

EARLY PHOTOGRAPH

An Oklahoma Geological Survey photograph taken before 1935 shows the parallel railroad and highway bridges on the North Fork of the Red River on the Kiowa-Greer County line near the town of Lugert. Many changes had occurred by 1990. Lake Lugert now covers the land where the town formerly stood. A concrete span had replaced the old steel-girder

BRIDGE ON HIGHWAY 44

MORE RECENT PHOTOGRAPH

bridge on OK-44, with well-grassed shoulders on the highway approaches to the bridge. Only the old railroad bridge remained apparently unchanged. American elm, black willow, southern hackberry, and salt cedar are tree species that grew in the bottomlands adjacent to the bridges and appear to be much larger and more numerous than in the 1935 photograph.

NUMBER FORTY-SEVEN

EARLY PHOTOGRAPH

Courtesy Oklahoma Geological Survey Collection, University of Oklahoma Libraries.

L. L. Hutchison took this photograph in 1909 of a distinctive mountain peak in Greer County one mile west of Quartz Mountain State Park. Nearly 90 years later, in 1995, the mountain and its vegetation remained much the same. Post oak-juniper savanna continued to dominate the rough granite slopes, while mesquite and redcedar covered the base of the mountain.

PEAK NEAR QUARTZ MOUNTAIN

MORE RECENT PHOTOGRAPH

The flatland on the approaches to the mountain was in cultivation at the time of the later photograph, replacing what appears to be grassland vegetation in the previous picture. This drastic change in the appearance of the prairie is in sharp contrast to the changeless profile of the distinctive granite mountain peak near the western edge of the Quartz Mountains in southwestern Oklahoma.

NUMBER FORTY-EIGHT

EARLY PHOTOGRAPH

Courtesy Oklahoma Geological Survey Collection, University of Oklahoma Libraries.

The town of Granite in Greer County is clearly visible in the distance in this 1935 panoramic view of western Oklahoma mountains and prairies. In the earlier picture there is little evidence of disturbance to the countryside as a result of human activity, except for the town itself. In 1990, when the later picture

TOWN OF GRANITE FROM WALSH MOUNTAIN

MORE RECENT PHOTOGRAPH

was made, the major changes are a fence line and the ground anchors of guy wires supporting a microwave relay tower on top of Walsh Mountain behind the site from which the picture was taken. Short grasses are typical plants in both pictures, along with hedgehog cactus and yucca growing where soil has accumulated between the rocks.

NORTHWEST OKLAHOMA

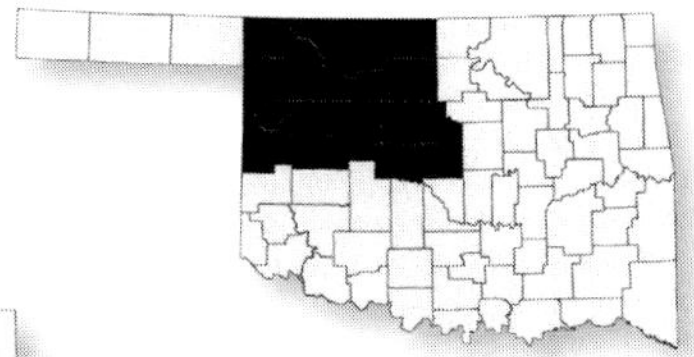

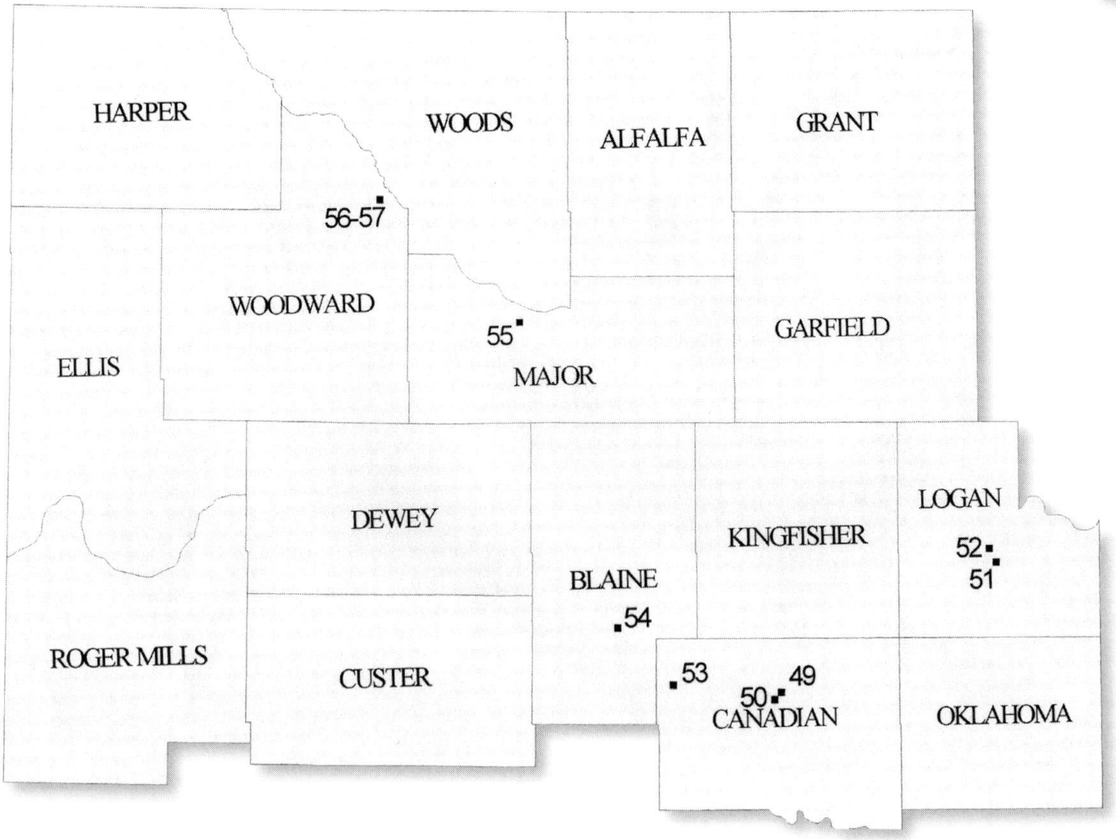

NUMBER FORTY-EIGHT

EARLY PHOTOGRAPH

Courtesy Shuck Collection, University of Oklahoma Libraries.

The Concho Indian School is the site of a pre-1935 photograph. Three men and two boys are resting on a wooden bridge spanning a creek bottom dominated by large trees. Grassy vegetation in the foreground and beneath the trees is typical of the prairie in the region.

CONCHO INDIAN SCHOOL

MORE RECENT PHOTOGRAPH

A winter scene at the site in 1990 shows a steel bridge with a wooden floor that had replaced the earlier wooden span. Tree species present in the creek bottom were hackberry, American elm, soapberry, eastern redcedar, and willow, along with several kinds of woody shrubs. A beaver dam in the creek had turned the stream into a shallow pool with a bed of cattails on the northern margin. Horseweed, ragweed, and other weedy species had become established in the grassy area on the slopes beside the stream. The bridge is across the road from the Cheyenne-Arapaho Agency at Concho in Canadian County.

NUMBER FIFTY

EARLY PHOTOGRAPH

Courtesy Shuck Collection, University of Oklahoma Libraries.

Courtesy Shuck Collection, University of Oklahoma Libraries.

Two early photographs record landscapes and vegetation at the Darlington Indian Agency near El Reno in Canadian County about 1890. One photograph, which includes a Native American encampment, shows tall grasses with a scattering of small trees and shrubs, characteristic of the North Canadian River bottom where the agency was located. The other photograph shows a scattering of horses and horsemen in the same area, and the grasses appear to have been reduced by grazing. Buildings, windmills, and water tanks of the agency are visible in the background of both photographs.

A third photograph in 1989 shows the open

DARLINGTON INDIAN AGENCY

MORE RECENT PHOTOGRAPH

field in cultivation where a Native American camp had stood a century before. Several barns and other structures of the agency which were visible in the 1890s are obscured by trees which had grown on the grounds. The old agency site belonged to the State of Oklahoma and was being used as a quail-breeding facility by the Oklahoma Department of Wildlife Conservation. In 1999, the game farm was abandoned and the land incorporated into the Oklahoma State University agricultural experiment station which adjoins it to the south.

NUMBER FIFTY-ONE

EARLY PHOTOGRAPH

Courtesy Western History Collection, University of Oklahoma Libraries.

A new bridge was under construction across Cottonwood Creek in Guthrie only two months after the town was founded in April, 1889. This photograph faces west on Noble Street and shows tents and wooden buildings in the creek floodplain. Steep banks more than 20 feet high mark the sides of the creek. Many large trees along the creek north and south of the bridge appear to be the cottonwoods that gave the creek its name. Tracks in the foreground were those of the Atchison, Topeka, and Santa Fe Railroad which was extended through the area in 1887.

A 1988 photograph shows an unusual two-level concrete bridge spanning Cottonwood Creek where the simpler wooden structure stood nearly a century before. The lower level

NOBLE STREET BRIDGE

MORE RECENT PHOTOGRAPH

of the newer bridge is about the height of the earlier one, and the higher one is above the level of the floods that still occur on the creek. The railroad tracks passing between the concrete supports of the upper level were still part of the Santa Fe system and seem to be in the same location as the 1889 tracks. Johnson grass was growing in the open spaces and along the banks of the creek. The creek bottom still supported numerous large native trees including American elm, red mulberry, hackberry, cottonwood, and black walnut, as well as tree-of-heaven, a widely planted Old World species that has escaped from cultivation.

NUMBER FIFTY-TWO

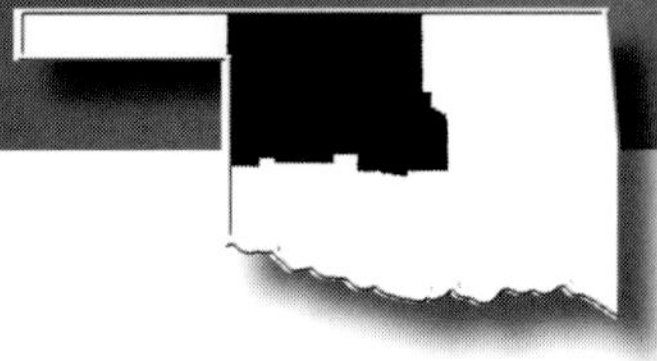

EARLY PHOTOGRAPH

Courtesy Oklahoma Geological Survey Collection, University of Oklahoma Libraries.

This 1916 photograph shows the Atchison Topeka & Santa Fe Railroad bridge on the Cimarron River in Logan County, north of Guthrie. The picture was taken from the highway bridge, the shadow of which can be seen in the lower right. The bare vertical bank at the right indicates undercutting and collapse of the sandy bank during spring floods. Riverbank vegetation appears to consist of a few trees in a low brushy growth.

RAILROAD BRIDGE ON THE CIMARRON

MORE RECENT PHOTOGRAPH

The 1988 photograph was taken from the US-77 bridge and shows a newer railroad bridge in the same location. Remains of bridge supports of two previous highway bridges can be seen in the foreground. The river bank vegetation appears to be much the same as in the older photograph and consists of a few trees such as soapberry, American elm, cottonwood, Chickasaw plum, sumac, eastern redcedar, and Johnson grass.

NUMBER FIFTY-THREE

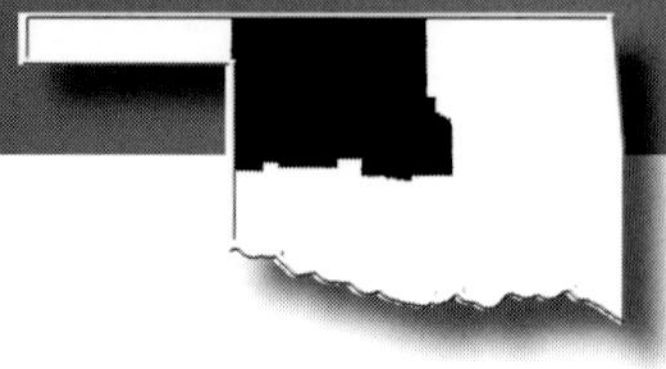

EARLY PHOTOGRAPH

Courtesy Oklahoma Geological Survey Collection, University of Oklahoma Libraries.

The flat-topped outline of Coyote Hill in Blaine County parallels the horizon in this 1900 photograph. The hill is a well-known local landmark visible from a county road about two miles east of Geary. It is the southern member of a pair of hills that lie in a north-south direction across the prairie. The top of Coyote Hill is slightly lower than its neighbor, from which these photographs were made.

COYOTE HILL

MORE RECENT PHOTOGRAPH

Vegetation on the hill had changed very little when the second photograph was made in 1988. On much of the prairie surrounding the hill, native short grasses that dominated in 1900 had been replaced by wheat fields and Bermuda grass. Native grasses, prickly pears, yucca, and locoweed still covered the slopes and top of the hill.

NUMBER FIFTY-FOUR

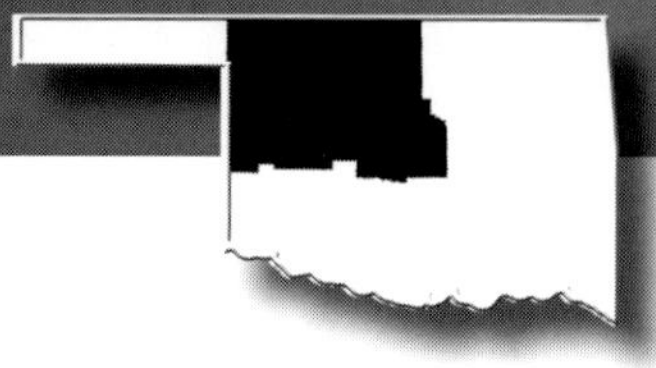

EARLY PHOTOGRAPH

Courtesy Phillips Collection, University of Oklahoma Libraries.

An undated photograph is described as "Red Hill, Blaine County." Original native prairie vegetation covers the flatland surrounding the hill. On the rocky slopes of the hill, similar prairie vegetation surrounds rocky outcrops and a few scattered clumps of eastern red cedar and other woody plants. The absence of a roadway and of other artifacts indicates the early date of the older picture.

RED HILL

MORE RECENT PHOTOGRAPH

US-281 now runs a few hundred yards east of the hill, 7.7 miles north of Geary. As can be seen in the 1997 photograph, the original prairie had been replaced by a farmyard and house near the base of the hill. A grove of honey locust trees and the house obscure much of the hill, although enough of the slope is visible to reveal an increase in the amount of woody vegetation on the hill.

NUMBER FIFTY-FIVE

EARLY PHOTOGRAPH

Courtesy Oklahoma Geological Survey Collection, University of Oklahoma Libraries.

Many flat-topped red buttes are found in the gypsum hills of Blaine and Major counties of northwestern Oklahoma. A distinctive example is Gypsum Butte, about 12 miles west of Orienta near OK-15. C. L. Cooper made the earlier photograph sometime before 1935.

GYPSUM BUTTE

MORE RECENT PHOTOGRAPH

The scene is almost unchanged in the 1989 photograph, except for improvements in the section line road and addition of a power line. Short grasses, mesquite, trees, and yucca are characteristic plants of the area. Herbaceous vegetation appears to be very sparse in the earlier picture which was probably made during the severe drought which occurred through much of the 1930s.

NUMBER FIFTY-SIX

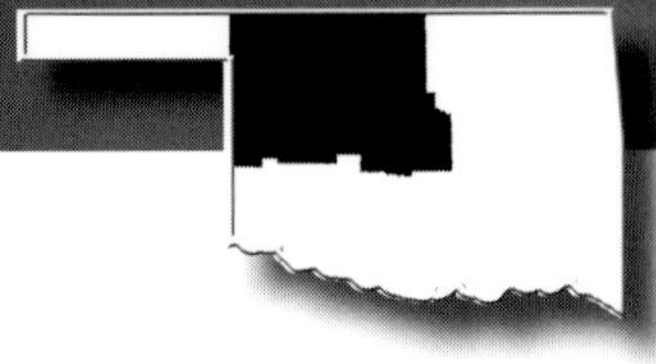

EARLY PHOTOGRAPH

Courtesy Oklahoma Geological Survey Collection, University of Oklahoma Libraries.

"Erosional chimney near mouth of Chimney Creek in Woodward County" is the original description of this undated photograph taken sometime before 1935. Before its collapse in the 1970s, the chimney was a prominent landmark among the buttes in this arid region in northwestern Oklahoma. The exact date of its collapse is unknown, but the distinctive column was no longer standing at the time of the 1989 photograph.

CHIMNEY ROCK

MORE RECENT PHOTOGRAPH

Vegetation in the surrounding lands consists of buffalo grass and other short grasses that are typical of the region, although more recently, a few shrubby mesquite trees, scattered forbs, and cacti had become established in the area. The chimney rock site is 19 miles north of the town of Orienta on US-60.

NUMBER FIFTY-SEVEN

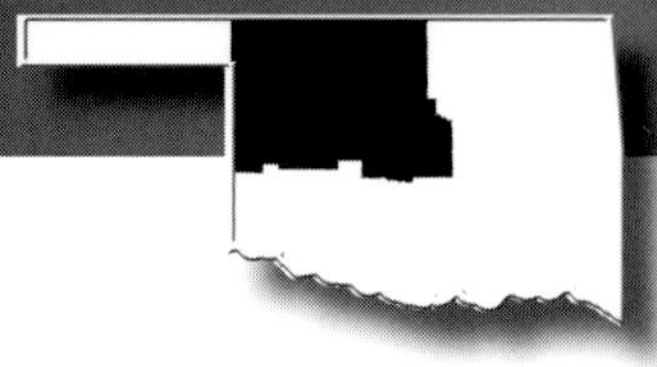

EARLY PHOTOGRAPH

Courtesy Oklahoma Geological Survey Collection, University of Oklahoma Libraries.

A more distant view of Chimney Rock and its surrounding gypsum escarpments shows changes over the decades. A cultivated field in the space between the camera site and the escarpment is visible in the early photograph and appeared to be unchanged in 1989.

DISTANT VIEW OF CHIMNEY ROCK

MORE RECENT PHOTOGRAPH

The invasion of mesquite into the area is evident as many of these shrubby trees can be seen along the nearside of the field. Grasses and forbs in the later photograph are similar to the species that occurred on the approaches to the escarpment closer to the base of the rock chimney.

THE PANHANDLE

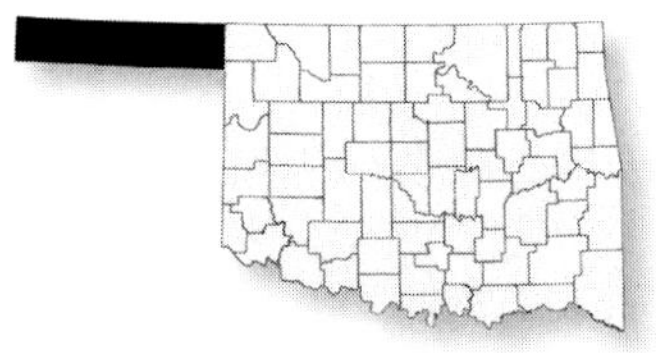

NUMBER FIFTY-EIGHT

EARLY PHOTOGRAPH

Courtesy Oklahoma Geological Survey Collection, University of Oklahoma Libraries.

Several bridges have spanned the Beaver River in Texas County north of Guymon on OK-3. The older photograph, made about 1935, shows one of the earlier bridges and gives an impression of the open, treeless nature of the floodplain in which it stood.

BEAVER RIVER BRIDGE

MORE RECENT PHOTOGRAPH

The Beaver River bridge site in 1988 had a markedly different appearance from that of a half century before. A new concrete span rises several feet above an older section which can be seen between trees in the left center of the newer photograph. The riverbanks had become thickly wooded with cottonwood, willow, and mulberry trees, obscuring all but short sections of the bridge. Switch grass and Johnson grass covered the sandy bottomland and salt cedar formed an under story beneath the larger trees.

NUMBER FIFTY-NINE

EARLY PHOTOGRAPH

Courtesy Stovall Collection, University of Oklahoma Libraries.

These photographs look west toward the mesa country near the extreme northwest corner of the Oklahoma Panhandle. The earlier photograph was made in 1937 by J. Willis Stovall of the University of Oklahoma. The site is called 101 Gap on the road formerly designated US-64 one mile east of the town of Kenton in Cimarron County.

101 GAP

MORE RECENT PHOTOGRAPH

The valley of the Cimarron River is beyond the mesa. Prominent woody vegetation in the older picture is cholla, yucca, snakeweed, and one-seeded juniper. The 1988 photograph reveals few changes in the appearance of the landscape. Forrest L. Johnson, co-author and research scientist with the Oklahoma Biological Survey, examines the vegetation on the roadside in the same location in the 1988 picture.

NUMBER SIXTY

EARLY PHOTOGRAPH

Courtesy Stovall Collection, University of Oklahoma Libraries.

A 1936 photograph shows several workers at this important paleontological excavation site in Cimarron County in the Oklahoma Panhandle. The pit is located on old US-64 7.5 miles east of the town of Kenton. Large numbers of dinosaur fossils were excavated from this site in the 1930s and removed for study to Stovall's laboratory at the University of Oklahoma.

STOVALL'S DINOSAUR QUARRY

MORE RECENT PHOTOGRAPH

The surrounding landscape consists of rolling short grass prairie on the rocky soil, with the horizon broken by the profile of one of several mesas in the vicinity. Little had changed in the appearance of the quarry and its surroundings in the half century. The pit had been slightly enlarged and a concrete replica of a dinosaur bone overlooked the quarry. Yucca and other small shrubs were more numerous than in 1936.

NUMBER SIXTY-ONE

EARLY PHOTOGRAPH

One of several balanced rocks along a ridge about seven miles east of Kenton appears in these two photographs of 1937 and 1988. The vegetation is short grass and appears to have been heavily grazed in 1937. A clump of taller grass, probably little bluestem, appears at the left side of the older photograph. Although the formation is composed of soft sandstone, no noticeable weathering had occurred in the 51 years between pictures.

Courtesy Stovall Collection, University of Oklahoma Libraries.

OLD MAID ROCK

MORE RECENT PHOTOGRAPH

The area appeared to be very lightly grazed in 1988 and, except for superficial differences caused by changes in grazing pressure, the vegetation appears to be unchanged. The short grasses visible in the more recent photograph are buffalo grass and blue grama. Toward the right side of the picture is sideoats grama. Little bluestem is the taller grass in the center.

NUMBER SIXTY-TWO

EARLY PHOTOGRAPH

Courtesy Stovall Collection, University of Oklahoma Libraries.

The subject of this 1937 photograph is Robbers' Roost Butte in Cimarron County, three miles north of the town of Kenton. The line of trees in the middle distance marks the channel of Carrizo Creek. Robbers' Roost is said to have been a hideout for the Dalton Gang during its run from the law in the late 1800s.

The upland vegetation on the butte is shrubby grassland. Trees along the creek bed appear to be cottonwoods. In the foreground is a mowed meadow or wheat field. A 1988 photo shows the upland vegetation on the butte to be unchanged. Grasses were gramas and buffalo grass. Snakeweed and one-seeded juniper were common. There are 123 junipers visible above the top of the first rock formation in the older

ROBBERS' ROOST

MORE RECENT PHOTOGRAPH

photograph and 131 in the same area of the newer photograph—thus little change had occurred on the upper slopes.

The number of junipers on the lower slopes of the butte between the lowest rock outcrop and the creek may have increased in the 51 years between pictures, but the view is obstructed by numerous large cottonwood trees making comparison between the older and more recent photograph impossible. Wheatgrass, blue grama, and buffalo grass had returned to the abandoned field visible in the foreground along with sand sage, cholla, and yucca.

NUMBER SIXTY-THREE

EARLY PHOTOGRAPH

Courtesy Stovall Collection, University of Oklahoma Libraries.

Tate butte is located five miles north of Kenton on Carrizo Creek Road. Vegetation on the butte and the lower upland to the left is shrubby grassland. A few junipers and small deciduous woody plants are on the colluvial slope to the left of the floodplain. Several large cottonwood trees stand at the base of the slope but almost no woody vegetation is visible on the banks of the creek in the 1937 photograph.

In 1988, vegetation on the butte and upland appeared virtually unchanged from the earlier time. Grasses were mostly grama and buffalo grass, with snakeweed the most common forb. One-seeded junipers visible on the mid and upper slopes of the butte are in the same locations in both photographs. Some are larger and are probably the same trees. This species is long-lived, although it remains small in size

TATE BUTTE AND CARRIZO CREEK

MORE RECENT PHOTOGRAPH

as older stems die and younger ones take their place.

A considerable increase in woody plant cover is noticeable on the cliff and colluvial slope in the center of the later view. Junipers, as well as deciduous shrubs, are visible in both pictures, but are larger in size and number in the later photograph. Deciduous woody plants growing on the cliff in 1988 included Virginia creeper, wild grape, aromatic sumac, chokecherry, raspberry, and wild currant. Only one large cottonwood from the earlier time remained in 1988, although the creek bank is now lined with young cottonwoods and salt cedars. Salt cedar, a recent invader, may not have been in the area at the time of the older photograph.

NUMBER SIXTY-FOUR

EARLY PHOTOGRAPH

Courtesy Oklahoma Geological Survey Collection, University of Oklahoma Libraries.

The older photograph shows Kenton as it appeared in 1916. The 1988 photograph reveals how Kenton has changed. In both photographs the camera is facing southeast in the direction of Main Street. In earlier times, the town was a center of trade for farmers in the Cimarron River Valley and ranchers in the surrounding uplands. Only two buildings visible in the original photo remained in 1988.

The ice cream parlor at the extreme right remained, but was no longer occupied. The first building visible to the right of the hitching rail shown in the older picture was the only commercial building in town in 1988 and served as post office, general store, and cafe. The decline of Kenton is typical of many small towns of Oklahoma and is a reflection of the overall decline in the rural agricultural population as

TOWN OF KENTON—LOOKING EAST

MORE RECENT PHOTOGRAPH

well as an increased mobility of the remaining residents.

Native vegetation on the mesa in the background is grama-buffalo short grass with scattered small woody and semi-woody plants such as cholla, yucca, and one-seeded juniper. The large trees visible in the older photograph are junipers. The short grass vegetation remained the same in 1988, but with some increase in the woody vegetation. There had been an obvious increase both in size and number of junipers on the mesa in the 72 years since the earlier picture was made. Thirty-eight junipers are visible in the earlier photograph, and 89 can be seen in a slightly smaller area in the newer one.

NUMBER SIXTY-FIVE

EARLY PHOTOGRAPH

Courtesy Oklahoma Geological Survey Collection, University of Oklahoma Libraries.

A view of Main Street looking west in Kenton is the subject of this photograph taken in February, 1928. The later photograph was made 60 years later. Native vegetation on the mesa in the background is grama-buffalo short grass, with scattered small woody and semi-woody plants such as cholla, yucca, and one-seeded juniper. The typical grama-buffalo short grass vegetation on the mesas in the background seemed unchanged in 1988, but there appeared to have been an increase both in size and number of junipers.

TOWN OF KENTON—LOOKING WEST

MORE RECENT PHOTOGRAPH

The presence of trees in the foreground in the newer picture made comparison difficult. The first two houses visible on the right in the older photograph were still standing at the time of the more recent picture. The first house constructed of native sandstone was unoccupied in 1988 and in poor repair. The date "1902" is carved into one of the stone blocks in the wall facing the street. The second house was occupied in 1988. Buildings visible on the left in the earlier picture were no longer standing.

NUMBER SIXTY-SIX

EARLY PHOTOGRAPH

Courtesy Stovall Collection, University of Oklahoma Libraries.

“Butte at bridge on Carrizo Creek, Oklahoma-New Mexico line. Cimarron County. 1937” reads the label on the older photograph. The view is facing southeast. Vegetation is shrubby grassland and appears to be the Shortgrass High Plains Type which is usually dominated by grama grasses and buffalo grass. At the time of the more recent photograph in 1988, snakeweed was the most common forb present, the same as 51 years before. The smaller, lighter-colored shrubs are cholla, yucca, and skunk bush . The larger, darker shrubs and small trees are one-seeded juniper.

The same species of plants appear to be present in both pictures. Herbaceous and small woody plants are in much the same condition

BUTTE WEST OF KENTON

MORE RECENT PHOTOGRAPH

and numbers in both pictures. The only noticeable difference is in the number and size of junipers. Several junipers visible on the mid- and upper slopes of the butte are in the same locations in both photos and appear larger in the newer, thus were probably the same trees. The greatest increase in numbers of junipers occurred on the nearly level ground between the observer and the base of the butte. Only nine junipers are visible below an arbitrary straight line drawn at the base of the butte in the older picture, while 21 are below the same line in the newer one. There are 91 visible above the line in the older photograph and 146 in the newer one.

NUMBER SIXTY-SEVEN

EARLY PHOTOGRAPH

Courtesy Forbes Collection, University of Oklahoma Libraries.

The older photograph is undated, but clothing styles and the sidesaddle on the ladies' patient steed indicate a date prior to 1920. The monument in the picture was erected in 1881 to mark the intersection of the 37th parallel and the 103rd meridian. Colorado was to the north, New Mexico Territory was to the west, and "No Man's Land," the Oklahoma Panhandle, lay to the east.

The survey was repeated in 1932 and a new monument erected about 300 yards to the south to mark the corrected location of the point of intersection of these two boundary lines. Shortgrass vegetation is on the flat land near the monument while a few junipers are visible on the higher land in the background in the old photograph. Although some changes are visible

COLORADO-NEW MEXICO-OKLAHOMA CORNER MARKER

MORE RECENT PHOTOGRAPH

in the 1988 photograph, the vegetation type was the same. Grasses were mostly grama, with a few patches of buffalo grass. Snakeweed and yucca are prominent in both photographs.

Quantitative changes appear in the vegetation which indicate deterioration in the condition of the range. Grass cover had decreased while bare ground and yucca had increased. The number of junipers on the hills and mesas in the background appears to have increased considerably. Cholla and aromatic sumac were common near the monument, although neither is visible in either photograph.

NUMBER SIXTY-EIGHT

EARLY PHOTOGRAPH

Courtesy Stovall Collection, University of Oklahoma Libraries.

The view is east from a point on the southeastern portion of the Black Mesa in Cimarron County in these two photographs made in 1941 and 1988. The Cimarron River is a prominent feature entering the picture from the right and vanishing into the distance in the center. Carrizo Creek appears as a nearly straight line of trees just left of center.

Vegetation on the flat upland areas both on top of the mesas and at the bottom of the slopes was the Shortgrass High Plains Type. A more complex type, a mixture of grassland and woody vegetation, occurred on the slopes. Woody vegetation along the river was primarily cottonwood. The more recent photograph was taken farther west and probably farther upslope than the earlier, but many features are common to both views.

CIMARRON RIVER VALLEY

MORE RECENT PHOTOGRAPH

Vegetation types were essentially the same as in 1941. The grassland on the mesa top and at the bottom of the slope was dominated by buffalo grass and grama with large numbers of chollas and yuccas. The complex vegetation on the cliffs and colluvial slopes was short grass with one-seeded juniper, aromatic sumac, cholla, and yucca in exposed locations. Areas which are more protected from sun and wind had mid-grasses, mostly little bluestem and sideoats grama, with netleaf hackberry and wafer ash. The river floodplain was dominated by cottonwood with patches of salt cedar along the riverbanks.

Upland vegetation visible in the more recent picture was very similar to that in the older one. Cottonwoods along the Cimarron River had expanded to cover most of the floodplain and form a continuous line along Carrizo Creek.

NUMBER SIXTY-NINE

EARLY PHOTOGRAPH

Courtesy Oklahoma Geological Survey Collection, University of Oklahoma Libraries.

The highest elevation in Oklahoma is 4,783 feet on the top of Black Mesa in Cimarron County, a few hundred yards east of the New Mexico state line. A rough monument of native rock capped with a carved block of granite was erected in the summer of 1928 to mark the spot. The older photograph was made while the monument was under construction. A few yuccas and small chollas are visible in an area of short grass.

HIGHEST POINT IN OKLAHOMA

MORE RECENT PHOTOGRAPH

A marker of polished granite was erected in 1954 at the place where its predecessor stood. The monument is located on a small area of state-owned land which has become part of Black Mesa State Park. In 1988, a rugged climb brought photographers to the same spot and inspired admiration for men and horses who 60 years earlier made the climb laden with building blocks of stone. Short grass seen in the later photograph appeared to be in much the same condition as in the earlier picture, but yuccas and chollas were more numerous and a few aromatic sumacs had become established. Most remarkable to find is goat's beard, an Old World introduction, growing in this remote and isolated location.

NUMBER SEVENTY

EARLY PHOTOGRAPH

Ruins of Fort Nichols are the only man-made feature visible on the high plains of western Cimarron County in these older undated photographs made sometime before 1935 by C. L. Cooper. No explanation is offered by Cooper for the difference in appearance of the fort in the two pictures.

The fort was constructed by Kit Carson in 1865 to protect travelers and traders on the Santa Fe Trail. It was abandoned after a few months although the trail remained in use until the railroad was built in the late 1880s. Buffalo grass, grama grass, and inconspicuous forbs are the typical vegetation of the short grass prairies through which the trail passes near the site of the old military installation. A few small shrubs, probably aromatic sumac, are visible on the rubble of the collapsed walls. The range appears

FORT NICHOLS

MORE RECENT PHOTOGRAPH

to be either heavily grazed or drought-stricken in the old photographs.

The ruins of Fort Nichols remained conspicuous marks on an otherwise pristine landscape at the time of the later 1988 photograph. The crumbling of the walls which is evident in the early photographs was then nearly complete. No portion of the ruins stood more than three feet high. A carpet of short grasses had softened the contour of the ruins, while one-seeded juniper and aromatic sumac had become sentries at intervals in the walls of the ruins of the fort. The short grasses were in very good condition in the summer of 1988 as a result of adequate rainfall and moderate grazing. Co-author Milby stands next to the rubble of the fort in the later photograph.

NUMBER SEVENTY-ONE

EARLY PHOTOGRAPH

Courtesy Oklahoma Geological Survey Collection, University of Oklahoma Libraries.

A monument was erected in 1903 at the southwestern corner of the Oklahoma Panhandle to mark the intersection of 36E 30' N and 103E W. The marker consists of a concrete-filled metal drum set in a concrete base. Lettering was done with a punch before the drum was filled with concrete.

The area was resurveyed in 1932 and a new monument placed about 100 yards to the southwest. This view is toward the northwest into New Mexico but the topography is similar in all directions from the site. Vegetation in the older photograph taken prior to 1932 is the Shortgrass High Plains Type, which is usually dominated by grama grasses and buffalo grass. No woody vegetation was growing in the area of

SOUTHWESTERN CORNER OF THE PANHANDLE

MORE RECENT PHOTOGRAPH

the marker.

Many changes had occurred in the vegetation of the area in the interval of time. In the 1988 photograph, a wheat field can be seen in the background with a center-pivot irrigation system visible on the horizon. Deep-well irrigation became a common practice in the panhandles of Oklahoma and Texas in the latter part of the 1900s, making it possible to grow corn and other cultivated crops on much of what was formerly the short grass prairies of the region. As a result of irrigation in some areas, such as this one surrounding the corner marker, only scattered patches of grama and buffalo grass remained of the original vegetation.

OKLAHOMA IN THE 21st CENTURY

Forecasting change is difficult, and predicting what changes will occur on Oklahoma landscapes in the future carries the risks that always come with speculation. It is nevertheless reasonable to make certain assumptions about forces that will shape the landscapes.

Climate will continue to influence landscapes and vegetation as it has done in the past. Extended periods of drought or of unusual rainfall may expand or reduce grassland areas and forests beyond their present boundaries. Fire or its suppression may continue as has been the case in the recent past to influence the invasion of eastern redcedar and other woody species into the grasslands of the state. The role that fire suppression has played is indicated by the increase from 1.5 million to 3.5 million acres of grassland invaded by redcedar in the decades between 1950 and 1985.

Among these factors, only the occurrence of fire is to any extent under human control. The extent to which fire is encouraged or suppressed will have a continuing influence on the character of vegetation in the years to come.

A typical scene showing a dense canopy of redcedar trees in a grassland in Cleveland County, one mile north of the town of Noble on US-77. Note the small trees in the foreground which have invaded the grassy area and are outside the perimeter of a recent fire. The larger trees were probably 30 to 40 years old when this photograph was made in 1997.

An example of urban sprawl.

However, it is the size and distribution of human population within the state, along with the nature of human activity, that may have the most profound effect on Oklahoma landscapes in the years ahead. Changes in the size and distribution of Oklahoma's population can be expected to reflect existing trends. Oklahoma's population will continue to increase although at a rate somewhat slower than the nation as a whole. The larger cities and towns will continue to grow while the population of many small towns and of rural parts of the state will decrease.

There will be a continuing urbanization of the state as is currently taking place. Urban sprawl and road construction will continue to alter the character of the countryside around cities and the larger towns. Farmland and open space will be converted to housing developments, parking areas, and commercial and industrial use on the periphery of urban areas.

Cultivated land may be converted to grassland, especially in the drier western parts of the state, while woodlands that had been cleared for cultivation in the eastern part of Oklahoma may be allowed to revert to forest.

Oklahoma has seen these kinds of changes in the past. The existence of Black Kettle National Grassland in Roger Mills County, which was established in the 1930s when native grasses reclaimed farmland that was abandoned because of drought, is evidence of the kinds of change that have occurred in the past and which the state may experience in the future.

More recently, the establishment in 1995 of Hackberry Flats Wildlife Management

Protracted drought throughout the Great Plains during the 1930s and the stranglehold of the Great Depression forced the abandonment of many farms in western Oklahoma. Much of this abandoned land was purchased by the United States government under the Industrial Recovery Act of 1933 to become the Black Kettle National Grassland. Photographs by Chuck Milner, Range Land Management Specialist at Black Kettle, show the region is predominately tall grass prairie broken by rolling hills and clusters of trees and other woody vegetation. A regime of controlled burns and grazing is used to restore and maintain the grassland vegetation. Although not within Black Kettle itself, an abandoned farmhouse is occasionally found nearby, a reminder of the farm life that existed in the area before economic and climatic conditions forced its demise.

area in Tillman County is an example of the continuing transformation of former agricultural landscapes to one approximating their natural aspect and vegetation. A marsh that had been drained and cultivated was acquired by the State of Oklahoma when farming was no longer practical. The land is being restored to its previous wetland condition and its natural vegetation will serve as a wildlife refuge and public hunting area in the future.

Roger Mills and Tillman counties, the sites of Black Kettle and of Hackberry Flats respectively, are both counties in which population has decreased in recent years. Existence of the refuge and the national grassland are example of changes in landscapes that have occurred as a result of shifts in populations in Oklahoma.

Similar kinds of changes in landscapes can be expected to occur in Oklahoma in the 21st century. These photographic records of Oklahoma landscapes of the past and of those taken in the latter part of the twentieth century provide insights into the ways Oklahoma has changed during the first century of settlement, and yet in many ways, has remained unchanged. The photographs also will provide a basis for study and reflection on an Oklahoma of the future when a similar photographic journey is made.

Hackberry Flats in Tillman County is being restored to its previous wetland condition by the State of Oklahoma.

Bibliography

Abert, J. W. *Journal of Lt. J. W. Abert from Bent's Fort to St. Louis in 1845*, U. S. 29th Cong., 1st Sess., Sen. Doc.438: 1-75 Congressional Serial Set 477, 1846.

Baird, W. David, and Danney Goble. *The Story of Oklahoma*. Norman: University of Oklahoma Press, 1994.

Bidwell, T. G. *Eastern Redcedar Ecology and Management*. OSU Extension Facts. No. 2868, Oklahoma State University, 1971.

Buck, S. J., *The Settlement of Oklahoma, Wisconsin Academy of Science, Arts and Letters*, XV (ii):325-380, 1907.

Boone, Nathan "Captain Nathan Boone's Journal," *Chronicles of Oklahoma*, 7:58-105, 1929.

Burke, Bob and John Boul. *Words of Destiny: The Writings that Shaped Oklahoma*. Oklahoma City: Oklahoma Historical Society, 2007.

Dale, Edward Everett, and Jesse Lee Rader. *Readings in Oklahoma History*. Evanston, Illinois: Row, Peterson and Co.,1939.

Foreman, Grant. *Adventures on the Red River: A Journal of Capt. R. B. Marcy*. Norman: University of Oklahoma Press, 1939.

Foreman, Grant. *Marcy and the Gold Seekers*. Norman: University of Oklahoma Press, 1939.

Gittinger, Roy. *The Formation of the State of Oklahoma (1803-1906)*. Berkely: University of California Press, 1917.

Goodman, G. J., and C. A. Lawson. *Retracing Major Stephen H. Long's 1820 expedition, the itinerary and botany*. Norman: University of Oklahoma Press, 1995.

Gould, C. N. "The dedication of the monument on Black Mesa, July 4, 1928," *Chronicles of Oklahoma* 7:34-54, 1929.

Gregg, Josiah, ed. by M. L. Moorhead. *Commerce of the Prairies.* Norman: University of Oklahoma Press, 1954.

Irving, Washington. *A tour of the prairie.* Norman: University of Oklahoma Press, 1956.

James, Edwin, 1823. *Account of an expedition from Pittsburgh to the Rocky Mountains performed in the years 1819, 1820. Vol. 2.* Philadelphia: Carey and Lea, 1823.

McReynolds, Edwin C. *Oklahoma, a history of the Sooner State.* Norman: University of Oklahoma Press, 1954.

Marcy, R. B. *Report of Capt. R. B. Marcy on the railroad from Fort Smith to Santa Fe.* 31C15 Ex. Doc. No. 43 Congressional Serial Set 577, 1850.

Nuttall, Thomas. *Journal of travels into the Arkansas Territory during the year 1819, with occasional observations on the manners of the aborigines.* Philadelphia: T.N. Palmer, 1821.

Nuttall, Thomas, ed. by Savoy Lottinville. *A journal of Travels into the Arkansas Territory during the Year 1819.* Norman: University of Oklahoma Press, 1980.

Thoburn, Joseph B. and Muriel H. Wright. *Oklahoma, a history of the state and its people.* New York: V. I. Lewis Historical Publishing Co., 1929.

Tomer, J. S., and M. J. Brodhead. *A Naturalist in Indian Territory. The Journals of S. W. Woodhouse, 1849-50.* Norman: University of Oklahoma Press, 1992.

Torrey, J. "Some account of a collection of plants made during a journey to and from the Rocky Mountains in the summer of 1820, by Edwin James, M. D., Assistant Surgeon, U. S. Army." *Annals of the Lyceum of Natural History of New York* , 2:161-254, 1828.

Wheelock, T. *Journal of the expedition of the regiment of dragoons into Indian country in 1834.* 23rd. Cong., 2nd. Sess., Sen. Doc. 1, No. 1:73-93, 1834.

PLANT NAMES USED IN THE TEXT

Common name	Scientific name
alder, seaside	*Alnus maritima*
ash, green	*Fraxinus pennsylvanica*
ash, wafer	*Ptelea trifoliata*
ash, white	*Fraxinus americana*
baldcypress	*Taxodium distichium*
basswood	*Tilia americana*
blackberry	*Rubus sp.*
blackhaw	*Viburnum rufidulum*
bluestem, little	*Schizachyrium scoparium*
bois d'arc	*Maclura pomifera*
boxelder	*Acer negundo*
cactus, hedgehog	*Echinocereus reichenbachii*
cactus, prickly pear	*Opuntia sp.*
catalpa	*Catalpa sp.*
cattail	*Typha sp.*
cedar, eastern red	*Juniperus virginiana*
cedar, rock	*Juniperus ashei*
chokecherry	*Prunus virginiana*
cholla	*Opuntia imbricata*
coreopsis, bigflower	*Coreopsis grandiflora*
cottonwood	*Populus deltoides*
creeper, virginia	*Parthenocissus quinquefolia*
currant, wild	*Ribes sp.*
elderberry	*Sambucus nigra ssp. canadensis*
elm, American	*Ulmus americana*
elm, winged	*Ulmus alata*
goat's beard	*Tragopogon sp.*
grama	*Bouteloua sp.*
grama, blue	*Bouteloua gracilis*
grama, sideoats	*Bouteloua curtipendula*
grape, wild	*Vitis sp.*
grass, Bermuda	*Cynodon dactylon*
grass, buffalo	*Buchloe dactyloides*
grass, Indian	*Sorghastrum nutans*
grass, Johnson	*Sorghum halepense*
greenbrier	*Smilax sp.*
gum, black	*Nyssa sylvatica*
gum, sweet	*Liquidambar styraciflua*
hackberry	*Celtis occidentalis*
hackberry, netleaf	*Celtis laevigata var. reticulate*
hackberry, southern	*Celtis laevigata*
hickory	*Carya sp.*
hickory, black	*Carya texana*
honeysuckle, Japanese	*Lonicera japonica*
horsetail	*Equisetum sp.*
ivy, poison	*Toxicodendron radicans*
juniper	*Juniperus sp.*
juniper, one-seeded	*Juniperus monosperma*
knotweed	*Polygonum sp.*
locoweed	*Oxytropis lambertii*
locust, black	*Robinia pseudoacacia*
locust, honey	*Gleditsia triacanthos*
maple, silver	*Acer saccharinum*
mesquite	*Prosopis glandulosa*
mulberry, red	*Morus rubra*
oak, Shumard	*Quercus shumardii*
oak, blackjack	*Quercus marilandica*
oak, bur	*Quercus macrocarpa*
oak, chinquapin	*Quercus muehlenbergii*
oak, post	*Quercus stellata*
oak, red	*Quercus rubra*
oak, short-lobed	*Quercus sinuata var. breviloba*
oak, water	*Quercus nigra*
pecan	*Carya illinoensis*
persimmon	*Diospyros virginiana*
pine, shortleaf	*Pinus echinata*
plum, Chickasaw	*Prunus angustifolia*
pokeweed	*Phytolacca americana*
raspberry	*Rubus sp.*
sage, sand	*Artemisia filifolia*
saltcedar	*Tamarix sp.*
skunkbush	*Rhus aromatica*
snakeweed	*Gutierrizia sarothrae*
soapberry	*Sapindus saponaria var. drummondii*
sumac, smooth	*Rhus glabra*
sumac, winged	*Rhus copallinum*
switchgrass	*Panicum virgatum*
sycamore	*Platanus occidentalis*
tree-of-heaven	*Ailanthus altissima*
verbena	*Verbena sp.*
walnut, black	*Juglans nigra*
water milfoil	*Myriophyllum sp.*
wheatgrass	*Pascopylum sp.*
willow	*Salix sp.*
willow, black	*Salix nigra*
willow, sandbar	*Salix exigua*
yucca	*Yucca sp.*

INDEX